Principles exist. We don't create them. We only discover them.

Vivekananda

An Introduction to Sieve Methods and Their Applications

LONDON MATHEMATICAL SOCIETY STUDENT TEXTS

Managing editor: Professor J. W. Bruce,
Department of Mathematics, University of Hull, UK

An Introduction to Sieve Methods and Their Applications

ALINA CARMEN COJOCARU
Princeton University

M. RAM MURTY
Queen's University

CAMBRIDGE
UNIVERSITY PRESS

CAMBRIDGE
UNIVERSITY PRESS

University Printing House, Cambridge CB2 8BS, United Kingdom

One Liberty Plaza, 20th Floor, New York, NY 10006, USA

477 Williamstown Road, Port Melbourne, VIC 3207, Australia

4843/24, 2nd Floor, Ansari Road, Daryaganj, Delhi - 110002, India

79 Anson Road, #06-04/06, Singapore 079906

Cambridge University Press is part of the University of Cambridge.

It furthers the University's mission by disseminating knowledge in the pursuit of education, learning and research at the highest international levels of excellence.

www.cambridge.org
Information on this title: www.cambridge.org/9780521612753

© Cambridge University Press 2005

First published 2006

A catalogue record for this publication is available from the British Library

ISBN 978-0-521-61275-3 Paperback

Contents

Preface

It is now nearly 100 years since the birth of modern sieve theory. The theory has had a remarkable development and has emerged as a powerful tool, not only in number theory, but in other branches of mathematics, as well. Until 20 years ago, three sieve methods, namely Brun's sieve, Selberg's sieve and the large sieve of Linnik, could be distinguished as the major pillars of the theory. But after the fundamental work of Deshouillers and Iwaniec in the 1980's, the theory has been linked to the theory of automorphic forms and the fusion is making significant advances in the field.

This monograph is the outgrowth of seminars and graduate courses given by us during the period 1995–2004 at McGill and Queen's Universities in Canada, and Princeton University in the US. Its singular purpose is to acquaint graduate students to the difficult, but extremely beautiful area, and enable them to apply these methods in their research. Hence we do not develop the detailed theory of each sieve method. Rather, we choose the most expedient route to introduce it and quickly indicate various applications. The reader may find in the literature more detailed and encyclopedic accounts of the theory (many of these are listed in the references). Our purpose here is didactic and we hope that many will find the treatment elegant and enjoyable.

Here are a few guidelines for the instructor. Chapters 1 through 5 along with Chapter 7 can be used as material for a senior level undergraduate course. Each chapter includes a good number of exercises suitable at this level. The book contains more than 200 exercises in all. Chapter 6 along with chapters 8 and 9 are certainly at the graduate level and require further prerequisites. Finally, Chapters 10 and 11 are at the 'seminar' level and require further mathematical sophistication. For the last chapter, in particular, a modest background in the theory of elliptic curves and automorphic representations may make the reading a bit smoother. Whenever possible, we have tried to provide suitable references for the reader for these prerequisites. Our list of references is by no means exhaustive.

1
Some basic notions

A. F. Möbius (1790–1868) introduced the famous Möbius function $\mu(\cdot)$ in 1831 and proved the now well-known inversion formula. The Möbius function is fundamental in sieve theory and it will be seen that almost all sieve techniques are based on approximations or identities of various sorts to this function.

It seems that the Möbius inversion formula was independently discovered by P. L. Chebycheff (1821–94) in the year 1851, when he wrote his celebrated paper establishing upper and lower bounds for $\pi(x)$ and settling Bertrand's postulate that there is always a prime number between n and $2n$ for $n \geq 2$. We give the proof of this below, but follow a derivation due to S. Ramanujan (1887–1920).

1.1 The big 'O' and little 'o' notation

Let D be a subset of the complex numbers $\mathbb{C}$ and let $f : D \longrightarrow \mathbb{C}$ be a complex valued map defined on D. We will write

$$f(x) = O(g(x))$$

if $g : D \longrightarrow \mathbb{R}^{+}$ and there is a positive constant A such that

$$|f(x)| \leq Ag(x)$$

for all $x \in D$. Often, D will be the set of natural numbers or the non-negative reals. Sometimes we will write

$$f(x) \ll g(x) \quad \text{or} \quad g(x) \gg f(x)$$

1

to indicate that $f(x) = O(g(x))$. If we have that $f(x) \ll g(x)$ and $g(x) \ll f(x)$, then we write

$$f(x) \asymp g(x).$$

In the case that D is unbounded, we will write

$$f(x) = o(g(x))$$

if

$$\lim_{\substack{x \to \infty \\ x \in D}} \frac{f(x)}{g(x)} = 0.$$

Clearly, if $f(x) = o(g(x))$, then $f(x) = O(g(x))$.

We will also write

$$f(x) \sim g(x)$$

to mean

$$\lim_{\substack{x \to \infty \\ x \in D}} \frac{f(x)}{g(x)} = 1.$$

As an example, look at $\nu(n)$, the number of distinct prime factors of a positive integer n. Then $\nu(n) = O(\log n)$. Also, for any $\varepsilon > 0$, $\log x = o\left(x^{\varepsilon}\right)$, because

$$\frac{\log x}{x^{\varepsilon}} \to 0$$

as $x \to \infty$.

There is one convention prevalent in analytic number theory and this refers to the use of ε. Usually, when we write $f(x) = O(x^{\varepsilon})$, it means that for any $\varepsilon > 0$ there is a positive constant C_{ε} depending only on ε such that $|f(x)| \leq C_{\varepsilon} x^{\varepsilon}$ for all $x \in D$. If the usage of ε is in this sense, then it should be clear that $f(x) = O\left(x^{2\varepsilon}\right)$ also implies $f(x) = O\left(x^{\varepsilon}\right)$.

Throughout the book, p, q, ℓ will usually denote primes, n, d, k will be positive integers, x, y, z positive real numbers. Any deviation will be clear from the context. Also, sometimes we denote gcd (n, d) as (n, d) and the lcm (n, d) as $[n, d]$.

1.2 The Möbius function

The **Möbius function**, denoted $\mu(\cdot)$, is defined as a multiplicative function satisfying $\mu(1) = 1$, $\mu(p) = -1$ for every prime p and $\mu(p^a) = 0$ for integers $a \geq 2$. Thus, if n is not squarefree, $\mu(n) = 0$, and if n is a product of k distinct primes, then $\mu(n) = (-1)^k$. We have the basic lemma:

Lemma 1.2.1 *(The fundamental property of the Möbius function)*

$$\sum_{d|n} \mu(d) = \begin{cases} 1 & \text{if } n = 1 \\ 0 & \text{otherwise.} \end{cases}$$

Proof If $n = 1$, then the statement of the lemma is clearly true. If $n > 1$, let $n = p_1^{a_1} \dots p_r^{a_r}$ be the unique factorization of n into distinct prime powers. Set $N = p_1 \dots p_r$ (this is called the **radical of** n). As $\mu(d) = 0$ unless d is squarefree, we have

$$\sum_{d|n} \mu(d) = \sum_{d|N} \mu(d).$$

The latter sum contains 2^r summands, each one corresponding to a subset of $\{p_1, ..., p_r\}$, since the divisors of N are in one-to-one correspondence with such subsets. The number of k element subsets is clearly

$$\binom{r}{k},$$

and for a divisor d determined by such a subset we have $\mu(d) = (-1)^k$. Thus

$$\sum_{d|N} \mu(d) = \sum_{k=0}^{r} \binom{r}{k}(-1)^k = (1-1)^r = 0.$$

This completes the proof. $\square$

Lemma 1.2.1 is fundamental for various reasons. First, it allows us to derive an inversion formula that is useful for combinatorial questions. Second, it is the basis of both the Eratosthenes' and Brun's sieves that we will meet in later chapters.

Theorem 1.2.2 *(The Möbius inversion formula)*
Let f and g be two complex valued functions defined on the natural numbers. If

$$f(n) = \sum_{d|n} g(d),$$

then

$$g(n) = \sum_{d|n} \mu(d) f(n/d),$$

and conversely.

Proof Exercise. $\square$

Theorem 1.2.3 *(The dual Möbius inversion formula)*

Let $\mathcal{D}$ be a divisor closed set of natural numbers (that is, if $d \in \mathcal{D}$ and $d'|d$, then $d' \in \mathcal{D}$). Let f and g be two complex valued functions on the natural numbers. If

$$f(n) = \sum_{\substack{n|d \\ d\in\mathcal{D}}} g(d),$$

then

$$g(n) = \sum_{\substack{n|d \\ d\in\mathcal{D}}} \mu\left(\frac{d}{n}\right) f(d),$$

and conversely (assuming that all the series are absolutely convergent).

Proof Exercise. $\square$

1.3 The technique of partial summation

Theorem 1.3.1 *Let $c_1, c_2, \ldots$ be a sequence of complex numbers and set*

$$S(x) := \sum_{n\leq x} c_n.$$

Let n_0 be a fixed positive integer. If $c_j = 0$ for $j < n_0$ and $f : [n_0, \infty) \longrightarrow \mathbb{C}$ has continuous derivative in $[n_0, \infty)$, then for x an integer $> n_0$ we have

$$\sum_{n\leq x} c_n f(n) = S(x)f(x) - \int_{n_0}^{x} S(t)f'(t)\mathrm{d}t.$$

Proof This is easily deduced by writing the left-hand side as

$$\sum_{n\leq x}\{S(n) - S(n-1)\}f(n) = \sum_{n\leq x} S(n)f(n) - \sum_{n\leq x-1} S(n)f(n+1)$$

$$= S(x)f(x) - \sum_{n\leq x-1} S(n)\int_{n}^{n+1} f'(t)\mathrm{d}t$$

$$= S(x)f(x) - \int_{n_0}^{x} S(t)f'(t)\mathrm{d}t,$$

because $S(t)$ is a step function that is constant on intervals of the form $[n, n+1)$. $\square$

In the mathematical literature, the phrase 'by partial summation' often refers to a use of the above lemma with appropriate choices of c_n and $f(t)$. For instance, we can apply it with $c_n = 1$ and $f(t) = \log t$ to deduce that

$$\sum_{n \leq x} \log n = [x] \log x - \int_1^x \frac{[t]}{t} dt,$$

where the notation $[x]$ indicates the greatest integer less than or equal to x. Since $[x] = x + O(1)$, we easily find:

Proposition 1.3.2

$$\sum_{n \leq x} \log n = x \log x - x + O(\log x).$$

Similarly, we deduce:

Proposition 1.3.3

$$\sum_{n \leq x} \frac{1}{n} = \log x + O(1).$$

1.4 Chebycheff's theorem

We caution the reader to what has now become standard notation in number theory. Usually p (and sometimes q or ℓ) will denote a prime number, and summations or products of the form

$$\sum_{p \leq x}, \quad \sum_p, \quad \prod_{p \leq x}, \quad \prod_p$$

indicate that the respective sums or products are over primes.

Let $\pi(x)$ denote the number of primes up to x. Clearly, $\pi(x) = O(x)$. In 1850, Chebycheff proved by an elementary method that

$$\pi(x) = O\left(\frac{x}{\log x}\right). \tag{1.1}$$

In fact, if we define

$$\theta(x) := \sum_{p \leq x} \log p,$$

then Chebycheff proved:

Theorem 1.4.1 *(Chebycheff's theorem)*
There exist positive constants A and B such that

$$Ax < \theta(x) < Bx.$$

By partial summation, this implies the bound on $\pi(x)$ stated in (1.1).

From Theorem 1.4.1 it is clear that there is always a prime number between x and Bx/A, since

$$\theta\left(\frac{Bx}{A}\right) > A\left(\frac{Bx}{A}\right) = Bx > \theta(x).$$

By obtaining constants A and B so that $B/A \leq 2$, Chebycheff was able to deduce further from this theorem:

Theorem 1.4.2 *(Bertrand's postulate)*
There is always a prime between n and 2n, for $n \geq 1$.

Chebycheff's theorem represented the first substantial progress at that time towards a famous conjecture of Gauss concerning the asymptotic behaviour of $\pi(x)$. Based on extensive numerical data and coherent heuristic reasoning, Gauss predicted that

$$\pi(x) \sim \frac{x}{\log x} \tag{1.2}$$

as $x \to \infty$. This was proven independently by Hadamard and de la Vallée Poussin in 1895 and is known as the **prime number theorem**. Before we discuss Chebycheff's proof, we will outline a simplified treatment of Theorem 1.4.1, due to Ramanujan [55].

Ramanujan's proof of Theorem 1.4.1 Let us observe that if

$$a_0 \geq a_1 \geq a_2 \geq \cdots$$

is a decreasing sequence of real numbers, tending to zero, then

$$a_0 - a_1 \leq \sum_{n=0}^{\infty}(-1)^n a_n \leq a_0 - a_1 + a_2.$$

These inequalities are obvious if we write

$$\sum_{n=0}^{\infty}(-1)^n a_n = a_0 - (a_1 - a_2) - (a_3 - a_4) - \cdots,$$

on the one hand, and

$$\sum_{n=0}^{\infty}(-1)^n a_n = (a_0 - a_1) + (a_2 - a_3) + \cdots,$$

on the other. Following Chebycheff, we define

$$\psi(x) := \sum_{p^a \le x} \log p, \qquad (1.3)$$

where the summation is over all prime powers $p^a \le x$. We let

$$T(x) := \sum_{n \le x} \psi\left(\frac{x}{n}\right)$$

and notice that

$$\sum_{n \le x} \log n = \sum_{n \le x} \left(\sum_{p^a | n} \log p\right) = \sum_{m \le x} \psi\left(\frac{x}{m}\right) = T(x).$$

By Proposition 1.3.2 of the previous section,

$$T(x) = x \log x - x + O(\log x),$$

so that

$$T(x) - 2T\left(\frac{x}{2}\right) = (\log 2)x + O(\log x).$$

On the other hand,

$$T(x) - 2T\left(\frac{x}{2}\right) = \sum_{n \le x}(-1)^{n-1}\psi\left(\frac{x}{n}\right)$$

and we can apply our initial observation to deduce

$$\psi(x) - \psi\left(\frac{x}{2}\right) + \psi\left(\frac{x}{3}\right) \ge (\log 2)x + O(\log x), \qquad (1.4)$$

because $a_n := \psi(x/n)$ is a decreasing sequence of real numbers tending to zero (in fact, equal to zero for $n > x$). By the same logic, we have

$$\psi(x) - \psi\left(\frac{x}{2}\right) \le (\log 2)x + O(\log x).$$

By successively replacing x with $x/2^k$ in the above inequality, we obtain

$$\psi(x) \le 2(\log 2)x + O\left(\log^2 x\right).$$

This completes Ramanujan's proof of Theorem 1.4.1. $\square$

We can easily deduce Bertrand's postulate now.

8 *Some basic notions*

Proof of Theorem 1.4.2 From (1.4),

$$\psi(x) - \psi\left(\frac{x}{2}\right) \geq \frac{1}{3}(\log 2)x + O\left(\log^2 x\right). \tag{1.5}$$

This shows that there is always a prime power between n and $2n$ for n sufficiently large. We leave as an exercise to show that (1.5) implies that there is always a prime between n and $2n$ for n sufficiently large. $\square$

We now indicate Chebycheff's argument.

Chebysheff's proof of Theorem 1.4.1 The key observation is that

$$\prod_{n < p \leq 2n} p \,\Big|\, \binom{2n}{n}.$$

Since

$$\binom{2n}{n} \leq 2^{2n},$$

we find, upon taking logarithms,

$$\theta(2n) - \theta(n) \leq 2n \log 2.$$

By writing successively

$$\theta(n) - \theta\left(\frac{n}{2}\right) \leq n \log 2,$$

$$\theta\left(\frac{n}{2}\right) - \theta\left(\frac{n}{4}\right) \leq \frac{n}{2} \log 2,$$

$$\cdots \quad \cdots$$

and by summing the inequalities, we find

$$\theta(2n) \leq 4n \log 2.$$

In other words,

$$\theta(x) = O(x).$$

Hence

$$x \gg \theta(x) \geq \sum_{\sqrt{x} < p \leq x} \log p$$

$$\geq \frac{1}{2}(\log x)\left(\pi(x) - \pi(\sqrt{x})\right)$$

$$\geq \frac{1}{2}(\log x)\pi(x) + O(\sqrt{x}\log x),$$

so that $\pi(x) = O\left(x/\log x\right)$.

Chebycheff also proved that, for some positive constant c,

$$\pi(x) > \frac{cx}{\log x}.$$

We relegate a proof of this (which is different from Chebycheff's) to an exercise. □

We will use the above circle of ideas to describe one more result of Chebycheff.

Theorem 1.4.3

$$\sum_{p \leq n} \frac{\log p}{p} = \log n + O(1).$$

Proof Let us study the prime factorization of $n!$. We write

$$n! = \prod_{p \leq n} p^{e_p},$$

since only primes $p \leq n$ can divide $n!$. The number of multiples of p that are $\leq n$ is $[n/p]$. The number of multiples of p^2 that are $\leq n$ is $\left[n/p^2\right]$, and so on. Hence it is easily seen that

$$e_p = \left[\frac{n}{p}\right] + \left[\frac{n}{p^2}\right] + \cdots,$$

where, in fact, the sum is finite since for some power p^a of p we will have $n < p^a$ so that $[n/p^a] = 0$. We therefore deduce

$$\log n! = \sum_{p \leq n} \left(\left[\frac{n}{p}\right] + \left[\frac{n}{p^2}\right] + \cdots \right) \log p.$$

Since we also have

$$\log n! = \sum_{k \leq n} \log k = n \log n - n + O(\log n)$$

and

$$\sum_{p \leq n} \left(\left[\frac{n}{p^2}\right] + \left[\frac{n}{p^3}\right] + \cdots \right) \log p \leq n \sum_p \frac{\log p}{p(p-1)} \ll n,$$

we find

$$\sum_{p \leq n} \left[\frac{n}{p}\right] \log p = n \log n + O(n).$$

□

Setting $c_n := (\log p)/p$ when $n = p$ and zero otherwise, we apply partial summation with $f(t) := (\log t)^{-1}$ to deduce, from Theorem 1.4.3:

Theorem 1.4.4

$$\sum_{p \leq n} \frac{1}{p} = \log \log n + O(1).$$

Remark 1.4.5 It is easy to see that in the above derivation we used Chebycheff's theorem only in getting an upper bound for the quantities in Theorems 1.4.3 and 1.4.4. Thus, the lower bound

$$\sum_{p \leq n} \frac{1}{p} \geq \log \log n + O(1)$$

can be obtained easily by partial summation from

$$\sum_{p \leq n} \frac{\log p}{p} \geq \log n + O(1),$$

which, in turn, can be derived without appealing to Chebycheff's upper bound (1.1).

1.5 Exercises

1. Show that $\log \log x = o((\log x)^\varepsilon)$ for any $\varepsilon > 0$.
2. If $d(n)$ denotes the number of divisors of n, show that $d(n) = O(\sqrt{n})$.
3. Show that there is a constant $c > 0$ such that
    $$d(n) = O\left(\exp\left(\frac{c \log n}{\log \log n}\right)\right).$$
    Deduce that for any $\varepsilon > 0$, $d(n) = O(n^\varepsilon)$.
4. Prove the Möbius inversion formula.
5. Prove the dual Möbius inversion formula.
6. Let F and G be complex valued functions defined on $[1, \infty)$. Prove that
    $$F(x) = \sum_{n \leq x} G(x/n)$$
    if and only if
    $$G(x) = \sum_{n \leq x} \mu(n) F(x/n).$$

7. Prove that there is a constant γ such that

$$\sum_{n \leq x} \frac{1}{n} = \log x + \gamma + O(1/x).$$

(γ is called **Euler's constant**.)

8. Show that

$$\sum_{n \leq x} d(n) = x \log x + O(x).$$

9. Prove that

$$\sum_{n \leq x} d(n) = x \log x + (2\gamma - 1)x + O(\sqrt{x}).$$

10. Prove, by partial summation, that $\psi(x) \sim x$ if and only if $\pi(x) \sim x/\log x$.
    [Hint: prove first that $\theta(x) \sim x$ if and only if $\pi(x) \sim x/\log x$.]

11. Using (1.4), show that

$$\theta(x) - \theta\left(\frac{x}{2}\right) \geq \frac{1}{3}(\log 2)x + O\left(x^{\frac{1}{2}} \log^2 x\right).$$

12. Define the function

$$\Lambda(n) := \begin{cases} \log p & \text{if } n = p^a \\ 0 & \text{otherwise.} \end{cases}$$

This is called **the von Mangoldt function**. Observe that

$$\psi(x) = \sum_{n \leq x} \Lambda(n)$$

and prove that

$$\psi(x) = O(x).$$

13. Show that

$$\sum_{n \leq x} \Lambda(n)^2 \ll x \log x.$$

14. Show that

$$e^{\psi(n)} = \mathrm{lcm}(1, 2, 3, \ldots, n).$$

15. Consider the integral

$$I = \int_0^1 x^n (1-x)^n \mathrm{d}x$$

and show that

$$0 < I < 4^{-n}.$$

16. Show that $e^{\psi(2n+1)} I$ is a positive integer.

17. Deduce from the previous exercise that $\psi(n) \geq (\log 2)n$ for n sufficiently large.

18. Infer from the previous exercise that for some constant $c > 0$,

$$\pi(x) > \frac{cx}{\log x}$$

for all $x \geq 2$.

19. Using the previous two exercises, show that there is always a prime between n and $2n$ for $n \geq 2$.

20. Let $\phi(n)$ denote the number of coprime residue classes mod n. This is called **Euler's function**. Show that

$$\sum_{d|n} \phi(d) = n.$$

21. Show that

$$\sum_{n \leq x} \frac{\phi(n)}{n} = \frac{6}{\pi^2} x + O(\log x).$$

22. Let p_n denote the n-th prime. Show that there are positive constants A and B such that $An \log n < p_n < Bn \log n$.

The following exercises utilise partial summation and Möbius inversion to deduce **Selberg's formula**. This was the key tool in Selberg's elementary proof of the prime number theorem [58], discovered in 1949.

23. Using partial summation, show that

$$\sum_{n \leq x} \log^2 n = x \log^2 x - 2x \log x + 2x + O\left(\log^2 x\right).$$

24. Using the previous exercise, deduce that

$$\sum_{n \leq x} \log^2 \frac{x}{n} = O(x).$$

25. Show that

$$\sum_{n \leq x} \frac{\Lambda(n)}{n} = \log x + O(1).$$

26. Show that

$$\sum_{n \leq x} \frac{\Lambda(n)}{n} \log \frac{x}{n} = \frac{1}{2} \log^2 x + O(\log x).$$

27. Putting $G(x) = 1$ in Exercise 6 above, deduce that

$$\sum_{n \le x} \frac{\mu(n)}{n} = O(1).$$

28. Putting $G(x) = x$ in Exercise 6 above, deduce that

$$\sum_{n \le x} \frac{\mu(n)}{n} \log \frac{x}{n} = O(1).$$

29. Putting $G(x) = x \log x$ in Exercise 6 above, deduce that

$$\sum_{n \le x} \frac{\mu(n)}{n} \log^2 \frac{x}{n} = 2 \log x + O(1).$$

30. Deduce from the previous exercises that

$$\sum_{d_1 d_2 \le x} \mu(d_1) \log^2 d_2 = 2x \log x + O(x).$$

31. A complex valued function defined on the set of natural numbers is called an **arithmetical function**. For two arithmetical functions f and g we define the **Dirichlet product** $f * g$ by

$$(f * g)(n) := \sum_{d|n} f(d)g(n/d).$$

We can also define the ordinary product of two arithmetical functions f and g, denoted fg, by setting $(fg)(n) := f(n)g(n)$. Now let

$$L(n) := \log n.$$

Show that

$$L(f * g) = (Lf) * g + f * (Lg).$$

32. If Λ denotes the usual von Mangoldt function, show that

$$\Lambda L + \Lambda * \Lambda = \mu * L^2,$$

where $L(\cdot)$ is as above.

33. Deduce from the previous exercises that

$$\sum_{n \le x} \Lambda(n) \log n + \sum_{uv \le x} \Lambda(u)\Lambda(v) = 2x \log x + O(x).$$

34. Deduce **Selberg's formula** from the previous exercise:

$$\psi(x) \log x + \sum_{n \le x} \Lambda(n)\psi(x/n) = 2x \log x + O(x).$$

35. Let $max(n)$ denote the largest exponent appearing in the unique factorization of n into distinct prime powers. Show that

$$\sum_{n \leq x} max(n) = O(x).$$

36. With $max(n)$ as in the previous exercise, show that for some constant $c > 0$,

$$\sum_{n \leq x} max(n) \sim cx$$

as x tends to infinity. What can be said about the error term? What can be said about the constant c?

2

Some elementary sieves

Modern sieve theory had an awkward and slow beginning in the early works of Viggo Brun (1885–1978). Brun's papers received scant attention and so sieve theory essentially lay dormant, waiting to be developed. By the 1950s, however, the Selberg sieve and the large sieve emerged as powerful tools in analytic number theory. In retrospect, we now understand the concept of a sieve in a better light. This reflection has recently given rise to some elementary sieve techniques. Though relatively recent in origin, these techniques are simple enough to be treated first, especially from the didactic perspective.

2.1 Generalities

Let $\mathcal{A}$ be a finite set of objects and let $\mathcal{P}$ be an index set of primes such that to each $p \in \mathcal{P}$ we have associated a subset $\mathcal{A}_p$ of $\mathcal{A}$. The **sieve problem** is to estimate, from above and below, the size of the set

$$\mathcal{S}(\mathcal{A}, \mathcal{P}) := \mathcal{A} \setminus \cup_{p \in \mathcal{P}} \mathcal{A}_p.$$

This is the formulation of the problem in the most general context. Of course, the 'explicit' answer is given by the familiar inclusion–exclusion principle in combinatorics. More precisely, for each subset I of $\mathcal{P}$, denote by

$$\mathcal{A}_I := \cap_{p \in I} \mathcal{A}_p.$$

Then the inclusion–exclusion principle gives us

$$\# \mathcal{S}(\mathcal{A}, \mathcal{P}) = \sum_{I \subseteq \mathcal{P}} (-1)^{\#I} \# \mathcal{A}_I,$$

where for the empty set $\emptyset$ we interpret $\mathcal{A}_\emptyset$ as $\mathcal{A}$ itself. This formula is the basis in many questions of probability theory (see the exercises).

15

In number theory we often take $\mathcal{A}$ to be a finite set of positive integers and $\mathcal{A}_p$ to be the subset of $\mathcal{A}$ consisting of elements lying in certain specified congruence classes modulo p. For instance, if $\mathcal{A}$ is the set of natural numbers $\leq x$ and $\mathcal{A}_p$ is the set of numbers in $\mathcal{A}$ divisible by p, then the size of $\mathcal{S}(\mathcal{A}, \mathcal{P})$ will be the number of positive integers $n \leq x$ coprime to all the elements of $\mathcal{P}$. Estimating $\#\mathcal{S}(\mathcal{A}, \mathcal{P})$ is a fundamental question which arises in many disguises in mathematics and forms the focus of attention of all sieve techniques. We will illustrate this in later chapters.

We could also reverse the perspective. Namely, we can think of $\mathcal{S} = \mathcal{S}(\mathcal{A}, \mathcal{P})$ as a given set, whose size we want to estimate. We seek to do this by looking at its image modulo primes $p \in \mathcal{P}$ for some set of primes $\mathcal{P}$. This will be the point of view in the large sieve (Chapter 8).

We could even enlarge this reversed perspective, as follows. Let $\mathcal{B}$ be a finite set of positive integers and let $\mathcal{T}$ be a set of prime powers. Suppose that we know the size of the image of $\mathcal{B}(\bmod t)$ for any $t \in \mathcal{T}$. We then seek to estimate the size of $\mathcal{B}$ itself. This is the approach of the larger sieve, to be discussed below. The rationale for the terminology will be explained later.

In some fortuitous circumstances, we may have a family $\mathcal{F}$ of complex valued functions

$$f : \mathcal{B} \longrightarrow \mathbb{C}$$

such that

$$\sum_{f \in \mathcal{F}} f(n) = \begin{cases} 1 & \text{if } n \in \mathcal{B} \\ 0 & \text{otherwise.} \end{cases} \tag{2.1}$$

Then

$$\#\mathcal{B} = \sum_{f \in \mathcal{F}} \left(\sum_{n \in \mathcal{B}} f(n) \right)$$

and the inner sum may be tractable by other techniques, such as analytic methods. We will illustrate this idea in the section on sieving using Dirichlet series.

Often it is the case that a precise relation as (2.1) may not exist and we may only have an approximation to it. For instance,

$$\sum_{f \in \mathcal{F}} f(n) = \begin{cases} 1 + o(1) & \text{if } n \in \mathcal{B} \\ o(1) & \text{otherwise,} \end{cases} \tag{2.2}$$

as $\#\mathcal{F} \to \infty$. Such a family of functions can then be used with great effect in obtaining estimates for $\#\mathcal{B}$.

Even the knowledge

$$\sum_{f \in \mathcal{F}} f(n) \geq 1 \quad \text{if } n \in \mathcal{B}$$

is sufficient to yield good upper bounds for $\#\mathcal{B}$. Indeed, we can consider

$$\#\mathcal{B} \leq \sum_{n \in \mathcal{B}} \left| \sum_{f \in \mathcal{F}} f(n) \right|^2$$

and expand out the square. After interchanging summations, we confront

$$\sum_{n \in \mathcal{B}} f(n) \overline{f'(n)}$$

for $f, f' \in \mathcal{F}$ (see Chapter 11).

We will illustrate these ideas, first with Gallagher's larger sieve [21] and then with the square sieve [24]. In the last section we will discuss analytic techniques alluded to earlier, namely how to sieve using Dirichlet series.

2.2 The larger sieve

Let $\mathcal{B}$ be a (non-empty) finite set of integers and let $\mathcal{T}$ be a set of prime powers. Suppose that for each $t \in \mathcal{T}$ we have

$$\#\mathcal{B}(\mathrm{mod}\ t) \leq u(t)$$

for some $u(t)$. Thus $\mathcal{B}$ represents at most $u(t)$ residue classes modulo t.

Theorem 2.2.1 *(Gallagher's larger sieve)*
We keep the above setting and let

$$X := \max_{b \in \mathcal{B}} |b|.$$

If

$$\sum_{t \in \mathcal{T}} \frac{\Lambda(t)}{u(t)} - \log(2X) > 0,$$

then

$$\#\mathcal{B} \leq \frac{\sum_{t \in \mathcal{T}} \Lambda(t) - \log(2X)}{\sum_{t \in \mathcal{T}} \dfrac{\Lambda(t)}{u(t)} - \log(2X)},$$

where $\Lambda(\cdot)$ is the von Mangoldt function.

Proof Let $t \in \mathcal{T}$ and for each residue class $r(\mathrm{mod}\ t)$ define

$$Z(\mathcal{B}; t, r) := \# \{b \in \mathcal{B} : b \equiv r(\mathrm{mod}\ t)\}.$$

Then

$$\#\mathcal{B} = \sum_{r(\mathrm{mod}\ t)} Z(\mathcal{B}; t, r).$$

By the Cauchy–Schwarz inequality, this is

$$\leq u(t)^{1/2} \left(\sum_{r(\mathrm{mod}\ t)} Z(\mathcal{B}; t, r)^2 \right)^{1/2}.$$

Hence

$$\frac{(\#\mathcal{B})^2}{u(t)} \leq \sum_{r(\mathrm{mod}\ t)} \sum_{\substack{b,b' \in \mathcal{B} \\ b,b' \equiv r(\mathrm{mod}\ t)}} 1$$

$$\leq \#\mathcal{B} + \sum_{\substack{b,b' \in \mathcal{B} \\ b \neq b'}} \sum_{t | b - b'} 1.$$

We multiply this inequality by $\Lambda(t)$ and we sum over $t \in \mathcal{T}$. Using

$$\sum_{t|n} \Lambda(t) = \log n,$$

we obtain

$$\sum_{t \in \mathcal{T}} \frac{(\#\mathcal{B})^2}{u(t)} \Lambda(t) \leq (\#\mathcal{B}) \sum_{t \in \mathcal{T}} \Lambda(t) + (\log 2X) \left((\#\mathcal{B})^2 - \#\mathcal{B} \right).$$

By cancelling the $\#\mathcal{B}$ and rearranging, we establish the inequality. $\square$

This sieve should be compared with the large sieve discussed in Chapter 8. The advantage here is that we can sieve out residue classes modulo prime powers, whereas in the large sieve only residue classes modulo primes are considered. This explains to some extent the name 'the larger sieve'.

Following Gallagher, we apply the larger sieve to prove:

Theorem 2.2.2 *Let a, b be integers having the property that for any prime power t there exists an integer v_t such that*

$$b \equiv a^{v_t}(\mathrm{mod}\ t).$$

Then there exists an integer v such that

$$b = a^v.$$

Before proceeding with the proof of this theorem, let us review some basic properties of the cyclotomic polynomial. Recall that for a positive integer d, the ***d*-th cyclotomic polynomial** $\Phi_d(x)$ is the minimal polynomial over $\mathbb{Q}$ of a primitive d-th root of unity. Thus it has degree $\phi(d)$, where $\phi(\cdot)$ is Euler's function. Now let n be an arbitrary positive integer. As the n-th roots of unity can be partitioned according to their order, we see that we have the formula

$$x^n - 1 = \prod_{d|n} \Phi_d(x).$$

Finally, for an integer a such that $(a, n) = 1$, let $f_a(n)$ be the order of a modulo n. By Exercise 6 we have that $f_a(n) = d$ if and only if $n | \Phi_d(a)$.

Proof of Theorem 2.2.2 Let a, b be as in the statement of the theorem. We note that to prove the result, we may suppose that a, b are positive and $a \geq 3$ (see Exercise 5).

Let

$$\mathcal{B} := \left\{ n \leq x : n = a^i b^j \text{ for some } i, j \right\}$$

and

$$\mathcal{T} := \left\{ t : t \text{ prime power}, f_a(t) \leq y \right\},$$

where $y = y(x)$ is some parameter to be chosen later. By Exercise 6, $\mathcal{T}$ is a finite set.

We keep the notation of Theorem 2.2.1. If for every prime power t we have that b is a power of a modulo t, then

$$u(t) \leq f_a(t).$$

Thus Theorem 2.2.1 implies that

$$\#\mathcal{B} \leq \frac{\displaystyle\sum_{t \in \mathcal{T}} \Lambda(t) - \log(2x)}{\displaystyle\sum_{t \in \mathcal{T}} \frac{\Lambda(t)}{f_a(t)} - \log(2x)}, \tag{2.3}$$

provided that the denominator is positive.

We have

$$\sum_{t \in \mathcal{T}} \Lambda(t) = \sum_{d \leq y} \sum_{f_a(t)=d} \Lambda(t)$$

$$= \sum_{d \leq y} \sum_{t | \Phi_d(a)} \Lambda(t)$$

$$= \sum_{d \leq y} \log \Phi_d(a),$$

upon using Exercise 6 and the formula $\sum_{d|n}\Lambda(d)=\log n$. Clearly,

$$(a-1)^{\phi(d)}\le|\Phi_d(a)|\le(a+1)^{\phi(d)},$$

so that

$$\sum_{t\in\mathcal{T}}\Lambda(t)=\sum_{d\le y}\log|\Phi_d(a)|\asymp\sum_{d\le y}\phi(d)\asymp y^2.$$

We also note that this implies

$$\sum_{t\in\mathcal{T}}\frac{\Lambda(t)}{f_a(t)}\ge\frac{1}{y}\sum_{t\in\mathcal{T}}\Lambda(t)\gg y.$$

Now choose

$$y:=100\log(2x).$$

From (2.3) we deduce that

$$\#\mathcal{B}\ll\log x.\qquad(2.4)$$

To this end, let us remark that if all the powers of a and b are distinct, then the set $\mathcal{B}$ has cardinality

$$\asymp(\log x)^2$$

(see Exercise 7). This contradicts (2.4), and so we conclude that for some i_0,j_0 we have

$$a^{i_0}=b^{j_0}.$$

We may even suppose that $(i_0,j_0)=1$, for otherwise we can take (i_0,j_0)-th roots of both sides of the above equality.

Let us write

$$n=\prod_p p^{\nu_p(n)}$$

for the unique factorization of an integer n into prime powers. We deduce that

$$i_0\nu_p(a)=j_0\nu_p(b)$$

for all primes p. As $(i_0,j_0)=1$, this means that $i_0|\nu_p(b)$ and $j_0|\nu_p(a)$ for all primes p. This implies that a is a j_0-th power and b is an i_0-th power of some integer c. The hypothesis now implies that for any prime q there exists a ν_q such that

$$c^{j_0\nu_q}\equiv c^{i_0}\pmod q,$$

which is equivalent to $f_c(q)|(j_0\nu_q-i_0)$ if $(q,c)=1$.

Now take a prime divisor q of $\Phi_{j_0 t}(c)$ for any t. By Exercise 6 we deduce that $f_c(q) \equiv 0 \pmod{j_0}$. Thus $j_0 | i_0$, and so b is a power of a, as desired. □

2.3 The square sieve

The square sieve is a simple technique originating in [24] meant to estimate the number of squares in a given set of integers. It relies on the use of a family of quadratic residue symbols for sifting out the squares. Consequently, it is well-suited for those sequences that are uniformly distributed in arithmetic progressions.

Theorem 2.3.1 *(The square sieve)*
Let $\mathcal{A}$ be a finite set of nonzero integers and let $\mathcal{P}$ be a set of odd primes. Set

$$S(\mathcal{A}) := \#\{\alpha \in \mathcal{A} : \alpha \text{ is a square}\}.$$

Then

$$S(\mathcal{A}) \leq \frac{\#\mathcal{A}}{\#\mathcal{P}} + \max_{\substack{q_1 \neq q_2 \\ q_1, q_2 \in \mathcal{P}}} \left| \sum_{\alpha \in \mathcal{A}} \left(\frac{\alpha}{q_1 q_2} \right) \right| + E,$$

where $\left(\frac{\cdot}{q_1 q_2} \right)$ denotes the Jacobi symbol and

$$E := O\left(\frac{1}{\#\mathcal{P}} \sum_{\alpha \in \mathcal{A}} \nu_{\mathcal{P}}(\alpha) + \frac{1}{(\#\mathcal{P})^2} \sum_{\alpha \in \mathcal{A}} \nu_{\mathcal{P}}(\alpha)^2 \right),$$

$$\nu_{\mathcal{P}}(\alpha) := \sum_{\substack{p \in \mathcal{P} \\ p | \alpha}} 1.$$

Remark 2.3.2 In practice, the contribution from E is negligible and one would expect the larger contributions to the estimate to come from the first two terms.

Proof We begin by observing that if $\alpha \in \mathcal{A}$ is a square, then

$$\sum_{q \in \mathcal{P}} \left(\frac{\alpha}{q} \right) = \#\mathcal{P} - \nu_{\mathcal{P}}(\alpha).$$

Thus

$$S(\mathcal{A}) \leq \sum_{\alpha \in \mathcal{A}} \frac{1}{(\#\mathcal{P})^2} \left(\sum_{q \in \mathcal{P}} \left(\frac{\alpha}{q} \right) + \nu_{\mathcal{P}}(\alpha) \right)^2. \tag{2.5}$$

Upon squaring and interchanging the summations, we get that the right hand side of inequality (2.5) is

$$\sum_{\alpha \in \mathcal{A}} \frac{1}{(\#\mathcal{P})^2} \left(\sum_{q_1, q_2 \in \mathcal{P}} \left(\frac{\alpha}{q_1} \right) \left(\frac{\alpha}{q_2} \right) + 2\nu_{\mathcal{P}}(\alpha) \sum_{q \in \mathcal{P}} \left(\frac{\alpha}{q} \right) + \nu_{\mathcal{P}}(\alpha)^2 \right).$$

The first sum is

$$\sum_{q_1, q_2 \in \mathcal{P}} \frac{1}{(\#\mathcal{P})^2} \sum_{\alpha \in \mathcal{A}} \left(\frac{\alpha}{q_1} \right) \left(\frac{\alpha}{q_2} \right) \leq \frac{\#\mathcal{A}}{\#\mathcal{P}} + \sum_{\substack{q_1, q_2 \in \mathcal{P} \\ q_1 \neq q_2}} \frac{1}{(\#\mathcal{P})^2} \sum_{\alpha \in \mathcal{A}} \left(\frac{\alpha}{q_1 q_2} \right)$$

$$\leq \frac{\#\mathcal{A}}{\#\mathcal{P}} + \max_{\substack{q_1, q_2 \in \mathcal{P} \\ q_1 \neq q_2}} \left| \sum_{\alpha \in \mathcal{A}} \left(\frac{\alpha}{q_1 q_2} \right) \right|.$$

The contribution to (2.5) from the latter sums is easily seen to be

$$E \leq \frac{2}{\#\mathcal{P}} \sum_{\alpha \in \mathcal{A}} \nu_{\mathcal{P}}(\alpha) + \frac{1}{(\#\mathcal{P})^2} \sum_{\alpha \in \mathcal{A}} \nu_{\mathcal{P}}(\alpha)^2.$$

This completes the proof. $\square$

Corollary 2.3.3 *Let $\mathcal{A}$ be a set of nonzero integers and let $\mathcal{P}$ be a set of primes that are coprime to the elements of $\mathcal{A}$. Then*

$$S(\mathcal{A}) = \#\{\alpha \in \mathcal{A} : \alpha \text{ is a square}\} \leq \frac{\#\mathcal{A}}{\#\mathcal{P}} + \max_{\substack{q_1, q_2 \in \mathcal{P} \\ q_1 \neq q_2}} \left| \sum_{\alpha \in \mathcal{A}} \left(\frac{\alpha}{q_1 q_2} \right) \right|.$$

Proof The hypothesis of the corollary implies that $\nu_{\mathcal{P}}(\alpha) = 0$ for any $\alpha \in \mathcal{A}$, so that $E = 0$ in the square sieve. $\square$

We want to apply the square sieve to count the number of integral points on a hyperelliptic curve

$$y^2 = f(x),$$

where $f(x) \in \mathbb{Z}[x]$ is a polynomial of degree d, of non-zero discriminant, and which is not the perfect square of a polynomial with integer coefficients. A famous theorem of Siegel [67] tells us that the number of integral points on such a curve is finite. Recently, effective estimates of this number have been given by various authors (see [30]). However, these estimates involve knowing the Mordell-Weil rank of the Jacobian of the hyperelliptic curve. Our approach is elementary and can be adapted to study how often a polynomial $f(x_1, \ldots, x_n)$ represents a square. As will be seen below, the generalization will require the deep work of Deligne (see [62]). For more applications, the reader may consult [7].

Given $f(x) \in \mathbb{Z}[x]$ and $k \in \mathbb{N}$, let us set for $k > 2$,

$$S_f(k) := \sum_{a(\mathrm{mod}\ k)} \left(\frac{f(a)}{k}\right),$$

where $(\cdot/k)$ is the Jacobi symbol.

Lemma 2.3.4 *Let q_1, q_2 be distinct primes and let $f \in \mathbb{Z}[x]$. Then*

$$S_f(q_1 q_2) = S_{f_1}(q_1) S_{f_2}(q_2),$$

where $f_1(x) := f(q_2 x), f_2(x) := f(q_1 x)$.

Proof The residue classes modulo $q_1 q_2$ can be written as

$$q_1 a_2 + q_2 a_1,$$

with $0 \leq a_2 \leq q_2 - 1, 0 \leq a_1 \leq q_1 - 1$ (see Exercise 8). Therefore

$$S_f(q_1 q_2) = \sum_{a_2=0}^{q_1-1} \sum_{a_1=0}^{q_2-1} \left(\frac{f(q_1 a_2 + q_2 a_1)}{q_1 q_2}\right)$$

$$= \sum_{a_2=0}^{q_1-1} \sum_{a_1=0}^{q_2-1} \left(\frac{f(q_1 a_2 + q_2 a_1)}{q_1}\right) \left(\frac{f(q_1 a_2 + q_2 a_1)}{q_2}\right)$$

$$= \sum_{a_2=0}^{q_1-1} \sum_{a_1=0}^{q_2-1} \left(\frac{f(q_2 a_1)}{q_1}\right) \left(\frac{f(q_1 a_2)}{q_2}\right).$$

The result follows. $\square$

Now let H be a positive real number and let us consider the set

$$\mathcal{A} := \{f(n) : |n| \leq H\}.$$

By the square sieve, the number of squares of $\mathcal{A}$ is, for any set of primes $\mathcal{P}$ not dividing the discriminant of f,

$$\leq \frac{2H+1}{\#\mathcal{P}} + \max_{\substack{q_1, q_2 \in \mathcal{P} \\ q_1 \neq q_2}} \left| \sum_{|n| \leq H} \left(\frac{f(n)}{q_1 q_2}\right) \right| + E,$$

where

$$E := O\left(\frac{H \log H}{\#\mathcal{P}} + \frac{H(\log H)^2}{(\#\mathcal{P})^2}\right)$$

and where we have used the elementary estimate $\nu_{\mathcal{P}}(\alpha) = O(\log \alpha)$.

Let q_1, q_2 be two distinct primes of $\mathcal{P}$. We have

$$\sum_{|n| \leq H} \left(\frac{f(n)}{q_1 q_2} \right) = \sum_{a \,(\mathrm{mod}\, q_1 q_2)} \left(\frac{f(a)}{q_1 q_2} \right) \sum_{\substack{|n| \leq H \\ n \equiv a \,(\mathrm{mod}\, q_1 q_2)}} 1.$$

The inner sum is

$$\frac{2H}{q_1 q_2} + O(1),$$

so we obtain

$$\sum_{|n| \leq H} \left(\frac{f(n)}{q_1 q_2} \right) = \frac{2H}{q_1 q_2} \sum_{a \,(\mathrm{mod}\, q_1 q_2)} \left(\frac{f(a)}{q_1 q_2} \right) + O(q_1 q_2).$$

By the lemma, the sum on the right-hand side is the product $S_{f_1}(q_1) S_{f_2}(q_2)$ for appropriate polynomials f_1, f_2.

We invoke a celebrated result of Weil (see [35, p. 99]), asserting that, for any $g(x) \in \mathbb{Z}[x]$ with non-zero discriminant and which is not the perfect square of a polynomial with integer coefficients, and for any prime p not dividing the discriminant of g,

$$\left| \sum_{a \,(\mathrm{mod}\, p)} \left(\frac{g(a)}{p} \right) \right| \leq (\deg g - 1) \sqrt{p}.$$

Using this in the above estimates gives

$$\sum_{|n| \leq H} \left(\frac{f(n)}{q_1 q_2} \right) = O\left(\frac{H}{\sqrt{q_1 q_2}} + q_1 q_2 \right).$$

Let us choose the set $\mathcal{P}$ to be given by the primes not dividing the discriminant of f and lying in the interval $[z, 2z]$ for some $z = z(H) > 0$, to be also chosen soon. We get the final estimate

$$S(\mathcal{A}) \ll \frac{H \log z}{z} + \frac{H}{z} + z^2 + \frac{H(\log H)(\log z)}{z} + \frac{H(\log H)^2(\log z)^2}{z^2}.$$

Choosing

$$z := H^{1/3} (\log H)^{2/3}$$

proves:

Theorem 2.3.5 *Let f be a polynomial with non-zero discriminant and integer coefficients, which is not the perfect square of a polynomial with integral coefficients. Let $H > 0$. Then the number of squares in the set*

$$\{ f(n) : |n| \leq H \}$$

is $O\left(H^{2/3}(\log H)^{4/3}\right)$, *with the implied O-constant depending only on the degree of* f *and the coefficients of* f.

2.4 Sieving using Dirichlet series

Sometimes, the sequences of numbers that we sift from exhibit a multiplicative structure and the sieve conditions may also exhibit such a property. In such cases, analytic methods using Dirichlet series are quite powerful and direct. In some instances, the techniques may even yield asymptotic formulae for the sieve problem. We illustrate this idea below, in greater detail. Further elaboration can be found in [46].

Let $\mathcal{P}$ be a set of primes and let $\overline{\mathcal{P}}$ indicate its complement in the set of all primes. Suppose that we want to count the number of natural numbers $n \leq x$ which are not divisible by any of the primes of $\mathcal{P}$. If we define the Dirichlet series

$$F(s) = \sum_{n \geq 1} \frac{a_n}{n^s} := \prod_{p \in \overline{\mathcal{P}}} \left(1 - \frac{1}{p^s}\right)^{-1},$$

we see that $a_n = 1$ if n is not divisible by any $p \in \mathcal{P}$ and $a_n = 0$ otherwise. Thus we seek to study

$$\sum_{n \leq x} a_n.$$

By Perron's formula (see [45, pp. 54–7]), this can be written as

$$\sum_{n \leq x} a_n = \frac{1}{2\pi i} \int_{2-i\infty}^{2+i\infty} F(s) \frac{x^s}{s} \, ds.$$

Here is a variant of the classical Tauberian theorem that is useful in such a context.

Theorem 2.4.1 *(Tauberian theorem)*
Let $F(s) = \sum_{n \geq 1} \dfrac{a_n}{n^s}$ *be a Dirichlet series with non-negative coefficients converging for* $\mathrm{Re}(s) > 1$. *Suppose that* $F(s)$ *extends analytically at all points on* $\mathrm{Re}(s) = 1$ *apart from* $s = 1$, *and that at* $s = 1$ *we can write*

$$F(s) = \frac{H(s)}{(s-1)^{1-\alpha}}$$

for some $\alpha \in \mathbb{R}$ *and some* $H(s)$ *holomorphic in the region* $\mathrm{Re}(s) \geq 1$ *and nonzero there. Then*

$$\sum_{n \leq x} a_n \sim \frac{cx}{(\log x)^\alpha}$$

with

$$c := \frac{H(1)}{\Gamma(1-\alpha)},$$

where Γ is the usual Gamma function.

This theorem was proven in 1938 by Raikov [54]. We do not prove it here, but only indicate that standard techniques of analytic number theory, as explained in [45, Chapter 4], can be used to derive the result. The reader may also find treatments in English in [75] and [68].

As an illustration of the principle, we consider the problem of counting the number of natural numbers $n \le x$ that can be written as the sum of two squares. It is well-known (see [32, p. 279]) that n can be written as a sum of two squares if and only if for every prime $p \equiv 3 \pmod 4$ dividing n, the power of p appearing in the unique factorization of n is even. Thus, if $a_n = 1$ whenever n can be written as a sum of two squares and is zero otherwise, we see that

$$F(s) := \sum_{n \ge 1} \frac{a_n}{n^s} = \left(1 - \frac{1}{2^s}\right)^{-1} \prod_{p \equiv 1 (\mathrm{mod}\ 4)} \left(1 - \frac{1}{p^s}\right)^{-1} \prod_{p \equiv 3 (\mathrm{mod}\ 4)} \left(1 - \frac{1}{p^{2s}}\right)^{-1}.$$

Now we need to invoke some basic properties of the **Riemann zeta function** $\zeta(s)$ and the **Dirichlet L-function** $L(s, \chi_4)$ associated with the quadratic character χ_4, defined by

$$\zeta(s) := \sum_{n \ge 1} \frac{1}{n^s}, \quad L(s, \chi_4) := \sum_{n \ge 1} \frac{\chi_4(n)}{n^s}$$

for $s \in \mathbb{C}$ with $\mathrm{Re}(s) > 1$. Here, $\chi_4(n)$ is 0 for n even and $(-1)^{(n-1)/2}$ for n odd. We refer the reader to [45] for the properties of these functions. Using the Euler products of $\zeta(s)$ and $L(s, \chi_4)$ we write

$$F(s) = [\zeta(s)L(s, \chi_4)]^{1/2} H_1(s),$$

where $H_1(s)$ is analytic and non-vanishing for $\mathrm{Re}(s) > 1/2$. As $L(s, \chi_4)$ extends to an entire function and is non-vanishing for $\mathrm{Re}(s) \ge 1$, we have

$$F(s) = \zeta(s)^{1/2} H_2(s)$$

for some $H_2(s)$ holomorphic and nonzero in $\mathrm{Re}(s) \ge 1$. Thus, using the fact that the Riemann zeta function has a simple pole at $s = 1$ and that it is analytic and non-vanishing for $\mathrm{Re}(s) = 1$, $s \ne 1$ (see [45]), we deduce that

$$F(s) = \frac{H(s)}{(s-1)^{1/2}},$$

with $H(s)$ holomorphic and non-vanishing in the region $\mathrm{Re}(s) \ge 1$.

By the Tauberian theorem cited above we obtain:

Theorem 2.4.2 *The number of $n \leq x$ that can be written as the sum of two squares is*

$$\sim \frac{cx}{\sqrt{\log x}}$$

for some $c > 0$, as $x \to \infty$.

By the same technique using classical Dirichlet series, one can also deduce the following remarkable result. Fix an integer $k \geq 3$. The number of $n \leq x$ not divisible by any prime $p \equiv 1 \pmod{k}$ is

$$\sim \frac{c_1 x}{(\log x)^{1/\phi(k)}}$$

for some $c_1 > 0$, where $\phi(k)$ is the Euler function. Thus a consequence of this result is that 'almost all' numbers have a prime divisor $p \equiv 1 \pmod{k}$.

Further applications of the technique can be found in [46].

2.5 Exercises

1. (The **Cauchy–Schwarz inequality**)

   Let $(a_i)_{1 \leq i \leq n}$, $(b_i)_{1 \leq i \leq n}$ be complex numbers. Show that

   $$\left| \sum_{1 \leq i \leq n} a_i b_i \right|^2 \leq \left(\sum_{1 \leq i \leq n} |a_i|^2 \right) \left(\sum_{1 \leq i \leq n} |b_i|^2 \right).$$

2. Let $k < n$ be positive integers. Let X be a k-element set and $Y = \{y_1, \ldots, y_n\}$ an n-element set. Let S be the set of all maps from X to Y, and $\mathcal{A}_i$ be the set of maps whose image does not contain y_i. Then the set

   $$S \setminus \cup_{1 \leq i \leq n} \mathcal{A}_i$$

   consists of maps that are surjective. Using the inclusion–exclusion principle, deduce that

   $$\sum_{0 \leq i \leq n} (-1)^i \binom{n}{i} (n-i)^k = 0$$

   whenever $k < n$.

3. Prove that

$$\sum_{0 \le i \le n} (-1)^i \binom{n}{i} (n-i)^n = n!.$$

4. Let $\mathcal{A} := \{1, 2, \ldots, n\}$. Denote by D_n the number of one-to-one maps $f : \mathcal{A} \longrightarrow \mathcal{A}$ without any fixed point. Show that

$$\lim_{n \to \infty} \frac{D_n}{n!} = \frac{1}{e},$$

    where e denotes Euler's e.

    [A map f without any fixed point is called a **derangement**].

5. Let a, b be integers. We say that a and b are related if, for every prime power t,

$$b \equiv a^{\nu_t} \pmod{t}$$

    for some positive integer ν_t. Show that if a and b are related, so are a^2 and b^2. Also, show that if $|a| \le 2$ and $|b| \le 2$ with a, b related, then $|a| = |b|$.

6. If t is a prime power coprime to k, show that $t | \Phi_k(a)$ if and only if $a \pmod{t}$ has order k.

7. If a and b are natural numbers ≥ 2 with $\{a^i : i \ge 1\} \cap \{b^j : j \ge 1\} = \emptyset$, show that the number of $a^i b^j \le x$ is

$$\asymp (\log x)^2.$$

8. Let t_1, t_2 be coprime positive integers and let $t := t_1 t_2$. Show that all the residue classes modulo t can be represented as

$$t_1 a_2 + t_2 a_1$$

    for some $0 \le a_2 \le t_2 - 1, 0 \le a_1 \le t_1 - 1$. Also, show that the coprime residue classes modulo t can be represented as above with $(a_2, t_2) = (a_1, t_1) = 1$.

9. For any $f(x) \in \mathbb{Z}[x]$ and any natural number $k > 2$, define

$$S_f(k) := \sum_{a \pmod{k}} \left(\frac{f(a)}{k} \right),$$

    where $(\cdot / k)$ is the Jacobi symbol. If $k = \prod_i q_i^{\alpha_i}$ is the unique prime factorization of k, then there exist polynomials f_i such that

$$S_f(k) = \prod_i S_{f_i}\left(q_i^{\alpha_i} \right).$$

10. A number n is called squareful if for every prime $p|n$ we have $p^2|n$. Show that the number of squareful natural numbers $\leq x$ is

$$\sim c_1\sqrt{x}$$

for some $c_1 > 0$, as $x \to \infty$.

11. Let k be a natural number ≥ 3. Show that the number of $n \leq x$ that are not divisible by any prime $p \equiv 1 \pmod{k}$ is

$$\sim \frac{c_2 x}{(\log x)^{1/\phi(k)}}$$

for some $c_2 > 0$, as $x \to \infty$.

12. Show that, for any positive integer n, $\sum_{d|n} \Lambda(d) = \log n$.

13. Show that

$$(a-1)^{\phi(d)} \leq |\Phi_d(a)| \leq (a+1)^{\phi(d)}$$

and

$$\sum_{d \leq x} \phi(d) \asymp x^2.$$

14. Let $\Psi(x, y)$ denote the number of $n \leq x$ with the property that if a prime $p|n$, then $p < y$. Show that if $x^{1/2} < z \leq x$, then

$$\Psi(x, z) = \left(1 - \log\left(\frac{\log x}{\log z}\right)\right)x + O\left(\frac{x}{\log z}\right).$$

[A number all of whose prime factors are $< y$ is called a y-**smooth number**.]

15. Prove that if $y < z$, then

$$\Psi(x, y) = \Psi(x, z) - \sum_{y \leq p < z}\sum_{r=1}^{\infty} \Psi(x/p^r, p).$$

Deduce that

$$\Psi(x, y) = \Psi(x, z) - \sum_{y \leq p < z} \Psi(x/p, p) + O(x/y).$$

[This is often referred to as **Buchstab's identity**.]

16. For $x^{1/3} \leq y \leq x^{1/2}$, show that

$$\Psi(x, y) \sim x\rho(u),$$

where $u := \log x / \log y$ and

$$\rho(u) := 1 - \log u + \int_2^u \log(v-1)\frac{dv}{v}.$$

[Hint: use Buchstab's identity and partial summation.]

17. Define $\rho(u)$ recursively by $\rho(u) := 1$ for $0 \le u \le 1$ and, for positive integers k, by

$$\rho(u) = \rho(k) - \int_k^u \rho(v-1)\frac{dv}{v}$$

for $k < u \le k+1$. Using Buchstab's identity, deduce inductively that for any $\varepsilon > 0$ and $x^\varepsilon < y \le x$, we have the asymptotic formula

$$\Psi(x, y) \sim x\rho(u)$$

with $u := \log x/\log y$. [$\rho(u)$ is called **Dickman's function** and was discovered by K. Dickman in 1930. For further details on this function, as well as more results concerning $\Psi(x, y)$, we refer the reader to [68, p. 367].]

18. Suppose that A is a subset of natural numbers contained in the interval $[1, x]$ whose image modulo every prime p has size $O(p^{1/2})$. Show that $\#A = O(\log x)$.

19. Let A be a set of natural numbers and let P be a set of prime numbers. Let (P) denote the semigroup generated by elements of P. Then any integer n can be written uniquely as $n = n_P m$, where $n_P \in (P)$ and m is coprime to p for all $p \in P$. Show that

$$\sum_{\substack{d|n \\ d\in(P)}} \mu(d)$$

equals 1 if $n_P = 1$ and zero otherwise. If $\Lambda_P(n)$ is defined to be $\log p$ whenever $n = p^a$ for some prime $p \in P$ and zero otherwise, then show that

$$\Lambda_P(n) = -\sum_{\substack{d|n \\ d\in(P)}} \mu(d)\log d$$

whenever $n_P > 1$.

20. With notation as in the previous exercise, let A be a set of natural numbers $\le x$ and let $S(A, P)$ denote the set of elements n of A with $n_P = 1$. Now suppose:

   (i) for $d \in (P)$, the set A_d consisting of elements of A divisible by d satisfies

$$\#A_d = \delta\frac{x}{d} + R(x, d)$$

for some δ and some $R(x, d)$ with

$$\sum_{\substack{d \leq x \\ d \in (P)}} |R(x, d)| \log \frac{x}{d} = O(x);$$

(ii) if $n \in A$ with $n_p > 1$, then n_p has at least two prime factors counted with multiplicity;

(iii) there is a set B such that $S(A, P) = S(B, P)$ and satisfying the condition that for $p \in P$ and $m \in B$, we have $pm \in B$;

(iv) there are numbers a and b with $a > 0$ so that

$$\sum_{\substack{m \leq x \\ m \in B}} \frac{1}{m} = a \log x + b + O\left(\frac{1}{x}\right).$$

Under these conditions, show that for some positive constant c,

$$\#S(A, P) \sim \frac{cx}{(\log x)^{1-a/b}}$$

as x tends to infinity.

21. Apply the previous exercise to the set A consisting of natural numbers $\equiv 1 \pmod 4$ and P the set of primes $\equiv 3 \pmod 4$. Thus, deduce Theorem 2.4.2. [Hint: take B to be the set of odd natural numbers.]

The last two exercises have been adapted from [52].

22. Let $f(x)$ be a polynomial with integer coefficients having the property that for every integer n, $f(n)$ is a perfect square. Show that $f(x)$ is the square of a polynomial with integer coefficients. Generalize this result to polynomials of several variables. [Hint: this can be deduced without using the results obtained in this chapter on the square sieve, as follows. We may suppose without loss of generality that $f(x)$ is a product of distinct irreducible polynomials. Take a prime p that is coprime to the discriminant of f such that $p|f(n)$ for some n. Consider $f(n+p)$ and $f(n)$ and deduce that one of them is not divisible by p^2. The general case can be formally treated using resultants and made to reduce to the single variable case.]

3

The normal order method

The normal order method has its origins in a 1916 paper of G. H. Hardy (1877–1947) and S. Ramanujan (1887–1920). A simpler and more transparent treatment of their work was given later in 1934 by Paul Turán (1910–76). Turán's method was substantially amplified by Paul Erdös (1913–96) and Mark Kac (1914–84). They used it to create an entire subject that has come to be known as probabilistic number theory.

The method of Turán will be discussed in greater detail in the next chapter, which will form the basis for an elementary sieve method. In this chapter we will focus on how Turán's method can be used to study the distribution of the number of prime factors of various sequences of numbers.

3.1 A theorem of Hardy and Ramanujan

We recall that $\nu(n)$ denotes the number of distinct prime divisors of n. In 1916, Hardy and Ramanujan (see [28, p. 356]) proved that almost all numbers n are composed of $\log \log n$ prime factors. To be precise, they showed that the number of $n \leq x$ not satisfying the inequality

$$(1 - \varepsilon) \log \log n < \nu(n) < (1 + \varepsilon) \log \log n$$

is $o(x)$ for any given $\varepsilon > 0$. Their proof involved an elaborate induction argument and was rather long and complicated. In 1934, Turán gave a simpler proof of their result. We begin by presenting Turán's proof. After that, we discuss the application of the technique to a wider context. In the next chapter, we will isolate a combinatorial sieve method from it, which we call the Turán sieve.

Theorem 3.1.1   *We have*

$$\sum_{n\leq x} \nu(n) = x \log \log x + O(x)$$

and

$$\sum_{n\leq x} \nu^2(n) = x(\log \log x)^2 + O(x \log \log x).$$

Proof   Let us observe that, by Theorem 1.4.4,

$$\sum_{n\leq x} \nu(n) = \sum_{p\leq x} \left[\frac{x}{p}\right]$$
$$= x \sum_{p\leq x} \frac{1}{p} + O(x)$$
$$= x \log \log x + O(x).$$

Also,

$$\sum_{n\leq x} \nu^2(n) = \sum_{n\leq x} \sum_{p|n} \sum_{q|n} 1$$
$$= \sum_{p,q\leq x} \sum_{\substack{n\leq x \\ p|n, q|n}} 1$$
$$= \sum_{\substack{p,q\leq x \\ p\neq q}} \left[\frac{x}{pq}\right] + \sum_{p\leq x} \left[\frac{x}{p}\right]$$
$$= \sum_{pq\leq x} \left[\frac{x}{pq}\right] + O(x \log \log x)$$
$$= x \sum_{pq\leq x} \frac{1}{pq} + O(x \log \log x).$$

Now,

$$\left(\sum_{p\leq\sqrt{x}} \frac{1}{p}\right)^2 \leq \sum_{pq\leq x} \frac{1}{pq} \leq \left(\sum_{p\leq x} \frac{1}{p}\right)^2.$$

Since

$$\sum_{p\leq\sqrt{x}} \frac{1}{p} = \log \log \sqrt{x} + O(1) = \log \log x + O(1),$$

we find that

$$\sum_{n\le x} \nu^2(n) = x(\log\log x)^2 + O(x\log\log x).$$

This completes the proof. $\square$

Now consider the variance

$$\sum_{n\le x}(\nu(n)-\log\log x)^2 = \sum_{n\le x}\nu^2(n) - 2(\log\log x)\sum_{n\le x}\nu(n) + (\log\log x)^2\sum_{n\le x}1,$$

which is easily seen to be $O(x\log\log x)$ by what we have proven above. This shows:

Theorem 3.1.2 *(Turán)*

$$\sum_{n\le x}(\nu(n)-\log\log x)^2 = O(x\log\log x).$$

Corollary 3.1.3 *Let $\delta > 0$. The number of $n \le x$ that do not satisfy the inequality*

$$|\nu(n)-\log\log x| < (\log\log x)^{\frac{1}{2}+\delta}$$

is $o(x)$.

Proof Indeed, if $n \le x$ does not satisfy the inequality, then a summand coming from n satisfies

$$|\nu(n)-\log\log x| \ge (\log\log x)^{1/2+\delta}.$$

The theorem implies that the number of such $n \le x$ is

$$O\left(x(\log\log x)^{-2\delta}\right) = o(x).$$

$\square$

Hardy and Ramanujan prove, in fact, a stronger theorem, getting a more precise estimate for the exceptional set. As already mentioned, the proof we have given is due to Turán (see, for example, [28, pp. 354–8]).

Related to the above results there is a celebrated theorem of Erdös and Kac (see [11, 12]), which states that, if for $\alpha \le \beta$,

$$S(x;\alpha,\beta) := \#\left\{n \le x : \alpha \le \frac{\nu(n)-\log\log n}{\sqrt{\log\log n}} \le \beta\right\},$$

then

$$\lim_{x \to \infty} \frac{S(x; \alpha, \beta)}{x} = \frac{1}{\sqrt{2\pi}} \int_\alpha^\beta e^{-t^2/2} dt.$$

The integral is the familiar probability integral associated with the normal distribution. Thus, the theorem of Erdös–Kac says that the function

$$\frac{\nu(n) - \log \log n}{\sqrt{\log \log n}}$$

is 'normally distributed' in a suitable probabilistic sense. We refer the reader to [11, 12] for further details.

We say that a function $f(n)$ has **normal order** $F(n)$ if, for every $\varepsilon > 0$, the inequality

$$(1 - \varepsilon)F(n) < f(n) < (1 + \varepsilon)F(n)$$

is satisfied for almost all values of n. That is, the number of $n \le x$ that do not satisfy the inequality is $o(x)$.

It is now not difficult to establish that $\nu(n)$ has normal order $\log \log n$ (see Exercise 1).

3.2 The normal number of prime divisors of a polynomial

Let $f(x)$ be an irreducible polynomial with integer coefficients. We would like to consider the problem of determining the normal order of $\nu(f(n))$. For this purpose, we proceed as in the previous section. The details follow.

First, let us observe that if $\nu_y(n)$ denotes the number of primes dividing n that are $\le y$ and if $y = x^\delta$ for some $0 < \delta < 1/2$, then for $n \le x$ we have

$$\nu(n) = \nu_y(n) + (\nu(n) - \nu_y(n)) = \nu_y(n) + O(1),$$

since the number of prime divisors of n greater than y is $O(1)$. We can therefore write

$$\sum_{n \le x} \nu(f(n)) = \sum_{n \le x} \nu_y(f(n)) + O(x). \qquad (3.1)$$

Let us denote by $\rho_f(p)$ the number of solutions modulo p of the congruence $f(x) \equiv 0 \pmod{p}$. Then

$$\sum_{n \le x} \nu_y(f(n)) = \sum_{\substack{n \le x \\ p \le y}} \sum_{p \mid f(n)} 1, \qquad (3.2)$$

so that, upon interchanging summation, we must count, for fixed p, the number of integers $n \le x$ that belong to $\rho_f(p)$ residue classes modulo p. We obtain

$$\sum_{n \le x} \nu(f(n)) = \sum_{p \le y} \left(\frac{x \rho_f(p)}{p} + O(\rho_f(p)) \right) + O(x), \tag{3.3}$$

and since $\rho_f(p) \le \deg f$, we see that the error term arising from the above summation is $O(y)$.

At this point we need to invoke some algebraic number theory. Let $K = \mathbb{Q}(\theta)$, where θ is a solution of $f(x) = 0$. The ring of integers $\mathcal{O}_K$ of K is a Dedekind domain. It is a classical theorem of Dedekind (see Theorem 5.5.1 of [17]) that for all but finitely many primes p, $\rho_f(p)$ is the number of prime ideals $\mathfrak{p}$ of $\mathcal{O}_K$ such that the norm $N_{K/\mathbb{Q}}(\mathfrak{p}) = p$. If $\pi_K(x)$ denotes the number of prime ideals whose norm is $\le x$, then the analogue of the prime number theorem for number fields asserts that

$$\pi_K(x) \sim \frac{x}{\log x}$$

as $x \to \infty$. In fact, for some constant $c > 0$,

$$\pi_K(x) = \operatorname{li} x + O(xe^{-c\sqrt{\log x}}),$$

where

$$\operatorname{li} x := \int_2^x \frac{dt}{\log t}$$

is the famous **logarithmic integral** (let us note that, upon integration by parts, $\operatorname{li} x = x/\log x + O\left(x/\log^2 x\right)$). Since the norm of any prime ideal is a prime power and since the number of prime ideals whose norm is not a prime cannot exceed $O(\sqrt{x}\log x)$, we deduce:

Theorem 3.2.1 *(The prime ideal theorem)*

$$\sum_{p \le x} \rho_f(p) = \operatorname{li} x + O(xe^{-c\sqrt{\log x}})$$

for some $c > 0$.

By partial summation we deduce further:

Corollary 3.2.2

$$\sum_{p \le x} \frac{\rho_f(p)}{p} = \log\log x + O(1).$$

We can now complete our analysis of the normal order of $\nu(f(n))$. By our earlier discussion ((3.1), (3.2), (3.3)) and the corollary above,

$$\sum_{n\leq x} \nu(f(n)) = x\log\log x + O(x).$$

Also,

$$\sum_{n\leq x} \nu^2(f(n)) = \sum_{n\leq x} \nu_y^2(f(n)) + O\left(\sum_{n\leq x} \nu_y(f(n))\right)$$

$$= \sum_{n\leq x} \nu_y^2(f(n)) + O(x\log\log x).$$

We find by the Chinese remainder theorem that

$$\sum_{n\leq x} \nu_y^2(f(n)) = \sum_{\substack{p,q\leq y \\ p\neq q}} \left(\frac{x\rho_f(p)\rho_f(q)}{pq} + O(1)\right) + O(x\log\log x),$$

where the latter error term arises from terms where $p = q$. Since

$$\sum_p \frac{\rho_f(p)^2}{p^2} = O(1),$$

we have

$$\sum_{\substack{p,q\leq y \\ p\neq q}} \frac{\rho_f(p)\rho_f(q)}{pq} = \left(\sum_{p\leq y} \frac{\rho_f(p)}{p}\right)^2 + O(1).$$

Thus

$$\sum_{n\leq x} \nu_y^2(f(n)) = x(\log\log x)^2 + O(y^2) + O(x\log\log x).$$

Now we recall that $y = x^\delta$ with $0 < \delta < 1/2$, and so the $O(y^2)$ error term is dominated by $O(x\log\log x)$. This proves:

Theorem 3.2.3

$$\sum_{n\leq x} (\nu(f(n)) - \log\log x)^2 = O(x\log\log x).$$

It is now an elementary exercise to deduce the normal order of $\nu(f(n))$ as $\log\log n$.

Theorem 3.2.3 gives an estimate of

$$\ll \frac{x}{\log\log x}$$

for the number of $n \leq x$ such that $f(n)$ is prime. A classical **conjecture of Buniakowski** formulated in 1854 predicts that any irreducible polynomial $f(x) \in \mathbb{Z}[x]$, such that $f(\mathbb{Z}^+)$ has no common divisor larger than 1, represents primes infinitely often. The only known case of this conjecture is the celebrated theorem of Dirichlet about the distribution of primes in an arithmetic progression, which settles it in the linear case. We will see later (using methods of Chapters 6 and 10) that sieve techniques can be applied to shed some light on this conjecture. In some cases, the methods come very close to settling it.

One can also establish the analogue of the Erdös–Kac theorem for $\nu(f(n))$. This has been done in [49].

If f is not irreducible, but has r irreducible factors, then the prime ideal theorem implies

$$\sum_{p \leq x} \rho_f(p) \sim \frac{rx}{\log x}$$

as $x \to \infty$. It will be of interest to make this effective and investigate whether such a result can be used to give an efficient 'primality test' or 'irreducibility test' for an arbitrary $f \in \mathbb{Z}[x]$.

It would also be of interest to generalize these investigations to study polynomials of several variables. This leads to the study of polynomial congruences modulo p in several variables, about which much is known thanks to the spectacular development of modern algebraic geometry (see, for example, the excellent monograph of Ireland and Rosen [32] for an introduction).

3.3 Prime estimates

We can investigate in a similar way $\nu(p-1)$ as p varies over the primes. More precisely, let k be a natural number and define for $(a, k) = 1$ the quantity

$$\pi(x; k, a) := \# \{ p \leq x : p \equiv a (\mathrm{mod}\ k) \} .$$

Then it is easily seen that

$$\sum_{p \leq x} \nu(p-1) = \sum_{\ell \leq x} \pi(x; \ell, 1),$$

where we recall that ℓ denotes a rational prime. As before, it is convenient to observe that, by an application of Chebycheff's theorem,

$$\sum_{p \leq x} \nu(p-1) = \sum_{p \leq x} \nu_y(p-1) + O\left(\frac{x}{\log x} \right),$$

so that we obtain sums of the form

$$\sum_{\ell \leq y} \pi(x; \ell, 1)$$

to investigate with $y = x^\delta$ for some $\delta > 0$.

A classical theorem of Bombieri and Vinogradov states that, for any $A > 0$, there is a $B = B(A) > 0$ so that

$$\sum_{k \leq x^{1/2} \log^{-B} x} \max_{y \leq x} \max_{(a,k)=1} \left| \pi(y; k, a) - \frac{\text{li } y}{\phi(k)} \right| \ll \frac{x}{\log^A x}.$$

This theorem will be proven in Chapter 9. In fact, much of the development of the large sieve method (to be discussed in Chapter 8) culminated in its proof. We invoke this theorem to deduce

$$\sum_{\ell \leq y} \pi(x; \ell, 1) = \sum_{\ell \leq y} \frac{\text{li } x}{\ell - 1} + O\left(\frac{x}{\log^A x}\right),$$

provided $\delta < 1/2$. Since

$$\sum_{\ell \leq y} \frac{1}{\ell - 1} = \sum_{\ell \leq y} \frac{1}{\ell} + O(1),$$

we conclude that

$$\sum_{p \leq x} \nu(p-1) = \pi(x) \log\log x + O\left(\frac{x}{\log x}\right).$$

In a similar way we deduce that (see Exercise 9)

$$\sum_{p \leq x} \nu^2(p-1) = \pi(x)(\log\log x)^2 + O(\pi(x)\log\log x).$$

This establishes the following theorem of Erdös:

Theorem 3.3.1 *(Erdös' theorem)*

$$\sum_{p \leq x} (\nu(p-1) - \log\log p)^2 = O\left(\frac{x \log\log x}{\log x}\right).$$

Again, it is possible to establish the analogue of the Erdös–Kac theorem for $\nu(p-1)$. In a recent paper [38], it is shown how one may prove this theorem without using the Bombieri–Vinogradov theorem but rather a 'weighted' version of the Turán sieve discussed in the next chapter.

A similar investigation can be made for the study of $\nu(p-a)$ for any integer a. The case $a = -2$ is related to the well-known **twin prime conjecture**. To be precise, a prime p is called a **twin prime** if $p+2$ is also a prime; it is conjectured that there are infinitely many such primes. An interesting corollary of the analogues of Theorem 3.3.1 for $\nu(p+2)$ will be that the number of primes $p \leq x$ such that $p+2$ is also a prime is

$$O\left(\frac{\pi(x)}{\log\log x}\right). \tag{3.4}$$

However, it is to be noted that we have proven (3.4) by assuming the Bombieri–Vinogradov theorem, which in turn is derived from the large sieve. The latter method is capable of yielding superior estimates for the number of twin primes, as will be discussed in Chapter 10.

3.4 Application of the method to other sequences

Having in mind the main steps for the proofs of Theorems 3.1.2, 3.2.3 and 3.3.1, we deduce that the **normal order method** can be formalized as follows.

Let $\mathcal{A} = (a_n)$ be a finite sequence of natural numbers. Let $\mathcal{A}_1 := \mathcal{A}$ and for each squarefree d, define

$$\mathcal{A}_d := \{a_n : a_n \equiv 0 (\mathrm{mod}\ d)\}.$$

For each squarefree d, we will write

$$\#\mathcal{A}_d = \frac{\omega(d)}{d}X + R_d,$$

or even

$$\#\mathcal{A}_d = \delta_d X + R_d,$$

where we think of X as an approximation to the cardinality of $\mathcal{A}$ and of R_1 as the error in the approximation. The function $\delta_d = \omega(d)/d$ is to be thought of as the 'proportion' of the elements of $\mathcal{A}$ belonging to $\mathcal{A}_d$. In particular, for primes p and q we write

$$\#\mathcal{A}_p = \frac{\omega(p)}{p}X + R_p$$

and

$$\#\mathcal{A}_{pq} = \frac{\omega(pq)}{pq}X + R_{pq}.$$

Now suppose that

$$a_n = O(n^C)$$

for some positive constant C. As before, we can show that

$$\sum_{n \le x} \nu(a_n) = \sum_{n \le x} \nu_y(a_n) + O(x)$$

for some $y = x^\delta$ and $\delta > 0$. Then we find

$$\sum_{n \le x} \nu(a_n) = \sum_{p \le y} \frac{\omega(p)}{p} X + \sum_{p \le y} R_p + O(x)$$

and we see that, in order to proceed further, we would need the asymptotic behaviour of

$$\sum_{p \le y} \frac{\omega(p)}{p}$$

and an estimate for the sum of the error terms

$$\sum_{p \le y} R_p.$$

Similarly, the study of

$$\sum_{n \le x} \nu^2(a_n)$$

would lead to finding an estimate for the sum

$$\sum_{p,q \le y} R_{pq}.$$

We will develop a general sieve out of this method in the next chapter.

In the example of Section 3.1, $a_n = n$, $X = x$ and $\omega(p) = 1$. In the example of Section 3.2, $a_n = f(n)$ with f an irreducible polynomial $\in \mathbb{Z}[x]$, $X = x$ and $\omega(p) = \rho_f(p)$. In the case $a_n = p_n - 1$ where p_n denotes the n-th prime and $p_n \le x$, we have $X = \pi(x)$ and $\omega(p) = p/(p-1)$.

There are many other interesting applications of this method. For example, let g be a nonzero **multiplicative function** (that is, $g(mn) = g(m)g(n)$ for any coprime integers m, n), taking rational integer values and such that $g(n) \ne 0$ for every natural number n (this assumption can be relaxed somewhat (see [48])). Define

$$\pi_g(x; d) := \#\{p \le x : g(p) \equiv 0 \pmod{d}\}.$$

Let us assume:
(H_0) for some $\gamma > 0$, $|g(n)| \le n^\gamma$ for all n;
(H_1) for some $\theta > 0$,

$$\sum_{d \le x^\theta} |\pi_g(x; d) - \delta(d)\pi(x)| \ll \frac{x}{\log x};$$

(H_2) for prime powers p^α, q^β $(p \neq q)$,

$$\delta(p^\alpha) = p^{-\alpha} + O(p^{-\alpha-1})$$

and

$$\delta(p^\alpha q^\beta) = (p^{-\alpha} + O(p^{-\alpha-1}))(q^{-\beta} + O(q^{-\beta-1})),$$

where the implied constants are absolute.

The following is proven in [49]:

Theorem 3.4.1 *Denote by*

$$N(x, \alpha) := \#\left\{ n \leq x : \frac{\nu(g(n)) - \frac{1}{2}(\log \log n)^2}{(\log \log n)^{3/2}} \leq \frac{\alpha}{\sqrt{3}} \right\},$$

where g satisfies $(H_0), (H_1), (H_2)$ *above. Then*

$$\lim_{x \to \infty} \frac{N(x, \alpha)}{x} = \frac{1}{\sqrt{2\pi}} \int_{-\infty}^{\alpha} e^{-t^2/2} dt.$$

Under the same hypotheses, one can also show that

$$\sum_{p \leq x} (\nu(g(p)) - \log \log p)^2 = O(\pi(x) \log \log x),$$

and, more generally, a theorem of Erdös–Kac type for $\nu(g(p))$. To be precise, if we let

$$S(x, \alpha) := \#\left\{ p \leq x : \frac{\nu(g(p)) - \log \log p}{\sqrt{\log \log p}} \leq \alpha \right\},$$

then

$$\lim_{x \to \infty} \frac{S(x, \alpha)}{\pi(x)} = \frac{1}{\sqrt{2\pi}} \int_{-\infty}^{\alpha} e^{-t^2/2} dt.$$

These results have been established in [48, 49].

Theorem 3.4.1 can, for instance, be applied to $\nu(\phi(n))$. In [16], Erdös and Pomerance determined the normal order of $\nu(\phi(n))$ using the Bombieri–Vinogradov theorem. In [50], Murty and Saidak show that the same result can be established without this theorem and by using only the elementary sieve of Eratosthenes (to be discussed in Chapter 5).

To cite a 'modular' example, consider the **Ramanujan function** $\tau(n)$ defined as the coefficient of x^n in the infinite product

$$x \prod_{m=1}^{\infty} (1 - x^m)^{24}.$$

By invoking the theory of ℓ-adic representations (see [63, 64]), one can prove certain properties about the number of prime divisors of $\tau(n)$. In 1916, Ramanujan conjectured that $\tau(n)$ is a multiplicative function. This was proven a year later by Mordell (see [61]). One can try to apply Theorem 3.4.1 to determine the normal number of prime factors of $\tau(n)$ whenever it is not zero. Such an investigation is caried out in [48, 49]. Hypothesis (H_0) is satisfied and hypotheses (H_1) and (H_2) are satisfied, as well, if we assume the generalized Riemann hypothesis for certain Artin L-functions. The method does lead to the interesting conclusion that

$$|\tau(p)| \geq \exp((\log p)^{1-\varepsilon})$$

for almost all primes p (that is, apart from $o(x/\log x)$ primes $p \leq x$) and for any $0 < \varepsilon < 1$. This is related to a classical **conjecture of Lehmer** (still unresolved) that $\tau(p) \neq 0$.

Finally, let us mention that the normal order method was recently applied by Ram Murty and F. Saidak to settle a conjecture of Erdös and Pomerance concerning the function $f_a(p)$, which is defined to be the number of prime factors of the order of a modulo p in $(\mathbb{Z}/p\mathbb{Z})^*$, albeit conditionally assuming the generalized Riemann hypothesis. Precise details can be found in [50].

3.5 Exercises

1. Prove that

$$\sum_{n \leq x} (\nu(n) - \log \log n)^2 = O(x \log \log x).$$

2. Let $\nu_y(n)$ denote the number of prime divisors of n that are less than or equal to y. Show that

$$\sum_{n \leq x} (\nu_y(n) - \log \log y)^2 = O(x \log \log y).$$

3. Prove that

$$\sum_{n \leq x} (\nu(n) - \log \log x)^2 = x \log \log x + O(x).$$

4. Let $\Omega(n)$ denote the number of prime powers that divide n. Show that $\Omega(n)$ has normal order $\log \log n$.

5. Fix $k \in \mathbb{Z}$ and let $(a, k) = 1$. Denote by $\nu(n; k, a)$ the number of prime divisors of n that are $\equiv a \pmod k$. Show that $\nu(n; k, a)$ has normal order

$$\frac{1}{\phi(k)} \log \log n.$$

6. Let g be a non-negative bounded function defined on the primes and define

$$g(n) := \sum_{p|n} g(p),$$

$$A(n) := \sum_{p \le n} \frac{g(p)}{p}.$$

Prove that

$$\sum_{n \le x} (g(n) - A(x))^2 = O(x \log \log x).$$

7. Using Theorem 3.2.1, prove that there is a positive constant c such that

$$\sum_{p \le x} \frac{\rho_f(p)}{p} = \log \log x + c + O\left(\frac{1}{\log x}\right).$$

8. If $f(x) \in \mathbb{Z}[x]$ is not irreducible, but has r distinct irreducible factors in $\mathbb{Q}[x]$ (hence in $\mathbb{Z}[x]$), deduce that $\nu(f(n))$ has normal order $r \log \log n$.

9. Using the Bombieri–Vinogradov theorem, prove that

$$\sum_{p \le x} \nu^2(p - 1) = \pi(x)(\log \log x)^2 + O(\pi(x) \log \log x)$$

10. Let P_x denote the greatest prime factor of

$$\prod_{n \le x} (n^2 + 1).$$

Show that $P_x > cx \log x$ for some positive constant c.

11. Show that, as $x \to \infty$,

$$\sum_{a^2 + b^2 \le x} \nu(a^2 + b^2) = \pi(x)(\log \log x) + O(x),$$

where the summation is over all integers a, b satisfying the inequality $a^2 + b^2 \le x$.

12. Let p and q be distinct primes. We write $p^k || n$ to mean that $p^k | n$ and p^{k+1} does not divide n. Show that the number of natural numbers $n \leq x$ such that $p^a || n$ and $q^b || n$ is less than or equal to

$$\frac{x}{p^a q^b}\left(1 - \frac{1}{p}\right)\left(1 - \frac{1}{q}\right) + 2.$$

13. Let f be an **additive arithmetical function**. Thus, by definition,

$$f(n) = \sum_{p^k || n} f(p^k)$$

where the summation is over prime powers p^k dividing n exactly. If f is a non-negative function, show that

$$\sum_{n \leq x} f(n) \geq x \sum_{p^a \leq x} \frac{f(p^a)}{p^a}\left(1 - \frac{1}{p}\right) - \sum_{p^a \leq x} f(p^a).$$

14. With f as in the previous exercise, show that

$$\sum_{n \leq x} f^2(n) \leq x E^2(x) + x V(x) + 2 \sum_{p^a q^b \leq x, p \neq q} f(p^a) f(q^b),$$

where

$$E(x) := \sum_{p^a \leq x} \frac{f(p^a)}{p^a}\left(1 - \frac{1}{p}\right)$$

and

$$V(x) := \sum_{p^a \leq x} \frac{f(p^a)^2}{p^a}.$$

15. With f, $E(x)$ and $V(x)$ as in the previous exercise, show that

$$\sum_{n \leq x} (f(n) - E(x))^2 \ll x V(x)$$

by expanding the square on the left hand side and using the inequalities obtained in the previous two exercises. This result is a special case of **the Turán–Kubilius inequality**. [Hint: observe that

$$\sum_{\substack{p^a q^b \leq x \\ p \neq q}} f(p^a) f(q^b) \leq \left(\sum_{p^a q^b \leq x} \frac{f(p^a)^2 f(q^b)^2}{p^a q^b}\right)^{1/2} \left(\sum_{\substack{p^a q^b \leq x \\ p \neq q}} p^a q^b\right)^{1/2}$$

by the Cauchy–Schwarz inequality.]

16. If f, $E(x)$ and $V(x)$ are as in the previous exercises and $V(x)$ tends to infinity as x tends to infinity, show that, for any $\varepsilon > 0$, the number of $n \le x$ with

$$|f(n) - E(x)| \ge V(x)^{1/2+\varepsilon}$$

   is $o(x)$.

17. Extend the result of Exercise 15 to arbitrary real valued additive functions f, as follows. Define additive functions $g_1(n)$ and $g_2(n)$ by setting $g_1(p^k) := f(p^k)$ if $f(p^k) \ge 0$ and zero otherwise, $g_2(p^k) := -f(p^k)$ if $f(p^k) < 0$ and zero otherwise. Then $f(n) = g_1(n) - g_2(n)$. If

$$E_j(x) := \sum_{p^k \le x} \frac{g_j(p^k)}{p^k} \left(1 - \frac{1}{p}\right),$$

   then show that

$$|f(n) - E(x)|^2 \le 2 \sum_{j=1}^{2} |g_j(n) - E_j(x)|^2$$

   by the Cauchy–Schwarz inequality. Deduce that

$$\sum_{n \le x} |f(n) - E(x)|^2 \ll xV(x).$$

18. Extend the result of the previous exercise to complex valued additive functions by considering the real-valued additive functions $g_1(n) := \mathrm{Re}(f(n))$ and $g_2(n) := \mathrm{Im}(f(n))$. This establishes the Turán–Kubilius inequality for all complex-valued additive functions.

19. Show that the implied constant in the Turán–Kubilius inequality can be taken to be 32.

20. Show that the factor $32x$ implied by the previous exercise for the Turán–Kubilius inequality can be replaced by

$$\Delta(x) := 2x + \left(\sum_{\substack{p^a q^b \le x \\ p \ne q}} p^a q^b\right)^{1/2} + 4 \left(\sum_{p^a \le x} \frac{1}{p^a} \sum_{q^b \le x} q^b\right)^{1/2}.$$

   Deduce that

$$\limsup_{x \to \infty} \frac{\Delta(x)}{x} \le 2.$$

4

The Turán sieve

In 1934, Paul Turán (1910–76) gave an extremely simple proof of the classical theorem of Hardy and Ramanujan about the normal number of prime factors of a given natural number. Inherent in his work is a basic sieve method, which was called Turán's sieve by the authors of [37], where it was first developed. In this chapter, we will illustrate how this sieve can be used to treat other questions that had previously been studied using more complicated sieve methods. For example, the Turán sieve is more elementary than the sieve of Eratosthenes and in some cases gives comparable results.

4.1 The basic inequality

Let $\mathcal{A}$ be an arbitrary finite set and let $\mathcal{P}$ be a set of prime numbers. For each prime $p \in \mathcal{P}$ we assume given a set $\mathcal{A}_p \subseteq \mathcal{A}$. Let $\mathcal{A}_1 := \mathcal{A}$ and for any squarefree integer d composed of primes of $\mathcal{P}$, let

$$\mathcal{A}_d := \cap_{p|d} \mathcal{A}_p.$$

Fix a positive real number z and set

$$P(z) := \prod_{\substack{p \in \mathcal{P} \\ p < z}} p.$$

We will be interested in estimating

$$S(\mathcal{A}, \mathcal{P}, z) := \# \left(\mathcal{A} \setminus \cup_{p|P(z)} \mathcal{A}_p \right).$$

Following the method illustrated in Section 3.4, we write for each prime $p \in \mathcal{P}$,

$$\#\mathcal{A}_p = \delta_p X + R_p \qquad (4.1)$$

47

and for distinct primes $p, q \in \mathcal{P}$,

$$\#\mathcal{A}_{pq} = \delta_p \delta_q X + R_{p,q}, \tag{4.2}$$

where

$$X := \#\mathcal{A},$$

$$0 \leq \delta_p < 1.$$

For notational convenience, we interpret $R_{p,p}$ as R_p. Heuristically, we usually think of δ_p as the proportion of elements of $\mathcal{A}$ lying in $\mathcal{A}_p$, and of R_p as the error term in this estimation. The same interpretation can be given to $\delta_p \delta_q$ and $R_{p,q}$.

Theorem 4.1.1   *(The Turán sieve)*

We keep the above setting. Let

$$U(z) := \sum_{p|P(z)} \delta_p.$$

Then

$$S(\mathcal{A}, \mathcal{P}, z) \leq \frac{X}{U(z)} + \frac{2}{U(z)} \sum_{p|P(z)} |R_p| + \frac{1}{U(z)^2} \sum_{p,q|P(z)} |R_{p,q}|.$$

Proof   For each element $a \in \mathcal{A}$, let $N(a)$ be the number of primes $p|P(z)$ such that $a \in \mathcal{A}_p$. Then

$$S(\mathcal{A}, \mathcal{P}, z) = \#\{a \in \mathcal{A} : N(a) = 0\} \leq \frac{1}{U(z)^2} \sum_{a \in \mathcal{A}} (N(a) - U(z))^2.$$

Thus the goal is to derive an upper bound for

$$\sum_{a \in \mathcal{A}} (N(a) - U(z))^2,$$

an expression that is reminiscent of the normal order method. Squaring out the summand and expanding, we must consider

$$\sum_{a \in \mathcal{A}} N(a)^2 - 2U(z) \sum_{a \in \mathcal{A}} N(a) + XU(z)^2.$$

For the first sum we have

$$
\sum_{a \in \mathcal{A}} N(a)^2 = \sum_{a \in \mathcal{A}} \left(\sum_{\substack{p|P(z) \\ a \in \mathcal{A}_p}} 1 \right)^2
$$

$$
= \sum_{p,q|P(z)} \# \mathcal{A}_p \cap \mathcal{A}_q
$$

$$
= \sum_{\substack{p,q|P(z) \\ p \neq q}} \# \mathcal{A}_{pq} + \sum_{p|P(z)} \# \mathcal{A}_p
$$

$$
= X \sum_{\substack{p,q|P(z) \\ p \neq q}} \delta_p \delta_q + X \sum_{p|P(z)} \delta_p + \sum_{p,q|P(z)} R_{p,q}
$$

$$
= X \left(\sum_{p|P(z)} \delta_p \right)^2 - X \sum_{p|P(z)} \delta_p^2 + X \sum_{p|P(z)} \delta_p + \sum_{p,q|P(z)} R_{p,q},
$$

and, similarly,

$$
\sum_{a \in \mathcal{A}} N(a) = X \sum_{p|P(z)} \delta_p + \sum_{p|P(z)} R_p.
$$

Here we have used assumptions (4.1) and (4.2). Therefore

$$
\sum_{a \in \mathcal{A}} (N(a) - U(z))^2 = X \sum_{p|P(z)} \delta_p (1 - \delta_p) + \sum_{p,q|P(z)} R_{p,q} - 2U(z) \sum_{p|P(z)} R_p.
$$

Since $(1 - \delta_p) \leq 1$, we immediately deduce the upper bound stated in the theorem. $\square$

Remark 4.1.2 In order to use the above theorem, the essential point is to have upper bounds for $R_{p,q}$ and R_p, as well as a lower bound for $U(z)$.

4.2 Counting irreducible polynomials in $\mathbb{F}_p[x]$

Let $\mathbb{F}_p$ denote the finite field of p elements. Fix a natural number $n > 1$ and let N_n be the number of monic irreducible polynomials in $\mathbb{F}_p[x]$ of degree n. There are several ways of obtaining an exact formula for N_n. The simplest is via a technique of zeta functions, as follows.

Consider the power series

$$
\sum_f T^{\deg f},
$$

where the summation is over all monic polynomials $f \in \mathbb{F}_p[x]$. Since the total number of monic polynomials $f \in \mathbb{F}_p[x]$ of degree n is p^n, the power series is easily seen to be

$$\sum_{n=0}^{\infty} p^n T^n = \frac{1}{1-pT}.$$

On the other hand, $\mathbb{F}_p[x]$ is a Euclidean domain and so it has unique factorization. Thus we can write an 'Euler product' expression for the power series above as

$$\prod_{v}(1 - T^{\deg v})^{-1} = \prod_{n=1}^{\infty}(1 - T^n)^{-N_n},$$

where v runs over monic irreducible polynomials of $\mathbb{F}_p[x]$. We therefore obtain

$$(1 - pT)^{-1} = \prod_{d=1}^{\infty}(1 - T^d)^{-N_d}. \tag{4.3}$$

By using that

$$-\log(1 - pT) = \sum_{n=1}^{\infty} \frac{p^n T^n}{n}$$

and taking logarithms in (4.3), we get

$$\sum_{n=1}^{\infty} \frac{p^n T^n}{n} = -\log(1 - pT) = -\sum_{d=1}^{\infty} N_d \log(1 - T^d)$$

$$= \sum_{d=1}^{\infty}\sum_{e=1}^{\infty} dN_d \frac{T^{de}}{de} = \sum_{n=1}^{\infty} \frac{T^n}{n}\left(\sum_{de=n} dN_d\right).$$

This proves:

Theorem 4.2.1 *Let N_d denote the number of monic irreducible polynomials of $\mathbb{F}_p[x]$ of degree d. Then*

$$\sum_{d\mid n} dN_d = p^n.$$

Observe that an immediate consequence is

$$N_n \le \frac{p^n}{n},$$

which can be viewed as the function field analogue of Chebycheff's upper bound (1.1) for $\pi(x)$. In fact, it is easy to deduce that (see Exercise 1)

$$N_n \sim \frac{p^n}{n},$$

or, more precisely,

$$N_n = \frac{p^n}{n} + O(p^{n/2}d(n)),$$

where $d(n)$ denotes the number of divisors of n. One can view this as the analogue of the prime number theorem (1.2) for $\mathbb{F}_p[x]$.

It is possible to invert the above expression for N_n and solve for it via the Möbius inversion formula. This will give us an explicit formula for N_n. More precisely, by applying Theorem 1.2.2 we deduce from Theorem 4.2.1:

Theorem 4.2.2 *Let N_n denote the number of monic irreducible polynomials of $\mathbb{F}_p[x]$ of degree n. Then*

$$N_n = \frac{1}{n} \sum_{d|n} \mu(d) p^{n/d}.$$

4.3 Counting irreducible polynomials in $\mathbb{Z}[x]$

Fix natural numbers H and $n > 1$. We will apply the Turán sieve to count the number of irreducible polynomials

$$x^n + a_{n-1}x^{n-1} + \cdots + a_1 x + a_0$$

with $0 \le a_i \le H$, $a_i \in \mathbb{Z}$. We will prove that this number is

$$H^n + O(H^{n-1/3} \log^{2/3} H).$$

Thus, a random polynomial with integer coefficients is irreducible, with probability 1.

Let us observe first that if a polynomial is reducible over $\mathbb{Z}[x]$, then it is reducible modulo p for every prime p. Our strategy will be to get an upper bound estimate for the number of reducible polynomials.

Let

$$\mathcal{A} := \{(a_{n-1}, a_{n-2}, \dots, a_1, a_0) \in \mathbb{Z}^n : 0 \le a_i < H\}.$$

We will think of the n-tuples $(a_{n-1}, \dots, a_1, a_0)$ as corresponding to the monic polynomials

$$x^n + a_{n-1}x^{n-1} + \cdots + a_1 x + a_0.$$

We want to count the number of tuples of $\mathcal{A}$ that correspond to irreducible polynomials in $\mathbb{Z}[x]$. So, let $\mathcal{P}$ consist of all primes and for each prime p, let $\mathcal{A}_p$ denote the subset of tuples corresponding to irreducible polynomials modulo p. Let $z = z(H)$ be a positive real number to be chosen later. Then $S(\mathcal{A}, \mathcal{P}, z)$ represents an upper bound for the number of *reducible* polynomials in $\mathbb{Z}[x]$, because if a polynomial belongs to $\mathcal{A}_p$ for some prime p, then it is irreducible.

We observe that $\mathcal{A}$ has H^n elements. If we specify a monic polynomial $g(x) \in \mathbb{F}_p[x]$, then the number of elements of $\mathcal{A}$ that, reduced modulo p, are congruent to $g(x) \pmod{p}$, is

$$\left(\frac{H}{p} + O(1) \right)^n.$$

We will choose z satisfying $z^2 < H$ so that, for primes $p < z$, this expression can be written as

$$\frac{H^n}{p^n} + O\left(\frac{H^{n-1}}{p^{n-1}} \right).$$

From our previous discussion, the number of monic irreducible polynomials of degree n is

$$N_n = \frac{p^n}{n} + O(p^{n/2}),$$

where the implied constant depends on n. Thus the total number of polynomials in $\mathcal{A}$ corresponding to irreducible polynomials of $\mathbb{F}_p[x]$ is

$$\left(\frac{H^n}{p^n} + O\left(\frac{H^{n-1}}{p^{n-1}} \right) \right) \left(\frac{p^n}{n} + O(p^{n/2}) \right) = \frac{H^n}{n} + O\left(\frac{H^n}{p^{n/2}} \right) + O(H^{n-1}p).$$

This implies that

$$\#\mathcal{A}_p = \frac{1}{n} H^n + O(H^{n-1}p) + O(H^n/p^{n/2})$$

and, similarly, that for $p \neq q$

$$\#\mathcal{A}_p \cap \mathcal{A}_q = \frac{1}{n^2} H^n + O(H^{n-1}pq) + O(H^n/p^{n/2}) + O(H^n/q^{n/2}).$$

We can now apply Theorem 4.1.1 with $\delta_p = 1/n$ and

$$R_{p,q} = O(H^{n-1}pq) + O(H^n/p^{n/2}) + O(H^n/q^{n/2}).$$

By using Chebycheff's bound, we deduce:

Theorem 4.3.1 *For $\mathcal{A}$ as above, $n \geq 3$ and $z^2 < H$, we have*

$$S(\mathcal{A}, \mathcal{P}, z) \ll \frac{H^n \log z}{z} + H^{n-1}z^2.$$

By choosing

$$z := H^{1/3}(\log H)^{1/3}$$

we obtain:

Theorem 4.3.2 *Let $n \geq 3$. The number of reducible polynomials*

$$x^n + a_{n-1}x^{n-1} + \cdots + a_1 x + a_0, \quad 0 \leq a_i \leq H, \ a_i \in \mathbb{Z},$$

is $O\left(H^{n-1/3}(\log H)^{2/3}\right)$.

For $n = 2$, the above analysis leads to an estimate of

$$O\left(H^{5/3}(\log H)^{2/3}(\log \log H)\right)$$

for the number of reducible monic quadratic polynomials, as the reader can easily verify. However, one can get a sharper estimate of $O(H \log H)$ for this, directly (see Exercise 6).

Gallagher [20] obtained the sharper estimate of $O(H^{n-1/2} \log H)$ by using a higher dimensional version of the large sieve inequality. One conjectures that the optimal exponent should be $n - 1$, and, by Exercise 6, this is best possible.

4.4 Square values of polynomials

Let $f(x) \in \mathbb{Z}[x]$ be a polynomial with non-zero discriminant disc f and which is not the square of another polynomial in $\mathbb{Z}[x]$. Let $H > 0$. We consider the question of estimating

$$\#\{|n| \leq H : f(n) \text{ is a perfect square}\}.$$

This question was discussed earlier in Chapter 2 in the context of the square sieve. We will now apply the Turán sieve by way of illustration of how such questions can be tackled by this sieve.

As noted earlier, Siegel's theorem concerning integral points on hyperelliptic curves gives us a universal bound (depending on f). The novelty here is that, on the one hand, we will not involve the deep work of Siegel. On the other hand, the method will have a wider range of applicability, such as the case of polynomials of several variables.

Let

$$\mathcal{A} := \{n : |n| \leq H\}$$

and for each prime $p \nmid \text{disc} f$, let

$$\mathcal{A}_p := \{|n| \leq H : f(n)(\text{mod } p) \text{ is not a square}\}.$$

By the result of Weil quoted in Chapter 2 (see [35, p. 94]),

$$\left| \sum_{a(\text{mod } p)} \left(\frac{f(a)}{p} \right) \right| \leq (\deg f - 1)\sqrt{p},$$

and so we find that the number of $n(\text{mod } p)$ such that

$$y^2 \equiv f(n)(\text{mod } p)$$

for some integer y is $p/2 + O\left(\sqrt{p}\right)$. Hence, the number of $n(\text{mod } p)$ so that $f(n)$ is not a square modulo p is $p/2 + O\left(\sqrt{p}\right)$. Thus

$$\#\mathcal{A}_p = \left(\frac{2H}{p} + O(1) \right) \left(\frac{p}{2} + O\left(\sqrt{p}\right) \right)$$

$$= H + O\left(\frac{H}{\sqrt{p}} + p \right).$$

Similarly, for distinct primes $p, q \nmid \text{disc } f$,

$$\#\mathcal{A}_{pq} = \left(\frac{2H}{pq} + O(1) \right) \left(\frac{pq}{4} + O\left(\sqrt{p}q + p\sqrt{q}\right) \right)$$

$$= \frac{H}{2} + O\left(pq + \frac{H}{\sqrt{p}} + \frac{H}{\sqrt{q}} \right).$$

As $\#\mathcal{A} = 2H + O(1)$, we see that, using the notation of Section 4.1, $\delta_p = 1/2$ and the axioms of the Turán sieve are satisfied with

$$R_p = O\left(\frac{H}{\sqrt{p}} + p \right)$$

and

$$R_{pq} = O\left(\frac{H}{\sqrt{p}} + \frac{H}{\sqrt{q}} + pq \right).$$

Applying Theorem 4.1.1 to our context, we obtain

$$\# \{|n| \leq H : f(n) \text{ is a square} \}$$

$$\ll \frac{H \log z}{z} + \frac{H}{\sqrt{z}} + z + \frac{1}{z^2} \left(Hz^{3/2} + z^4 (\log z)^2 \right).$$

Choosing $z := \frac{H^{2/5}}{(\log H)^{4/5}}$, we get

$$\# \{|n| \leq H : f(n) \text{ is a square} \} = O\left(H^{4/5} (\log H)^{2/5} \right),$$

which is inferior to the estimate obtained by the square sieve.

We note, however, that if we consider the analogous problem of how often $f(n)$ is a cube or a higher odd power, the square sieve is inapplicable and we leave it to the reader to show that a similar estimate can be deduced by the above technique.

4.5 An application with Hilbert symbols

Let $H > 0$. We now consider the problem of counting the number of pairs of integers (a, b) with $1 \leq a, b \leq H$ with the property that

$$ax^2 + by^2 = z^2$$

has a non-trivial rational point. We will need some properties of the Hilbert symbol. These will be reviewed here, but we also refer the reader to [61, Ch. 3] for a more detailed exposition.

Let k be $\mathbb{R}$ or the p-adic field $\mathbb{Q}_p$. For $a, b \in k^*$, we define the **Hilbert symbol** (a, b) to be equal to 1 if $ax^2 + by^2 = z^2$ has a non-trivial solution (i.e. different from $(0,0,0)$) and to be equal to -1 otherwise. The Hilbert symbol satisfies

$$(aa', b) = (a, b)(a', b).$$

If $k = \mathbb{Q}_p$ we may write $(a, b)_p$ for (a, b) to indicate the dependence on p and if $k = \mathbb{R}$ we may write $(a, b)_\infty$.

If $k = \mathbb{Q}_p$ with $p > 2$, the Hilbert symbol can be computed using Legendre symbols, as follows. Let $a = p^\alpha u$ and $b = p^\beta v$ with $u, v \in \mathbb{Z}_p^*$. Then

$$(a, b)_p = (-1)^{\frac{\alpha\beta(p-1)}{2}} \left(\frac{u}{p}\right)^\beta \left(\frac{v}{p}\right)^\alpha,$$

where (u/p) denotes the Legendre symbol $(\bar{u}/p)$, with $\bar{u}$ denoting the image of u under the homomorphism of reduction (modulo p).

By a classical theorem of Hasse and Minkowski (see [61, p. 41]), we have that

$$ax^2 + by^2 = z^2$$

has a non-trivial rational point if and only if

$$(a, b)_p = 1 \; \forall p \text{ and } (a, b)_\infty = 1.$$

Now let $\mathcal{P}$ be an arbitrary set of primes, let

$$\mathcal{A} := \{(a, b) : 1 \leq a, b \leq H\},$$

and for each prime $p \in \mathcal{P}$ let

$$\mathcal{A}_p := \left\{(a, b) \in \mathcal{A} : (a, b)_p = -1\right\}.$$

Then the size of

$$\mathcal{A} \backslash \bigcup_{p \in \mathcal{P}} \mathcal{A}_p$$

gives us an upper bound for the number of (a, b) with $1 \le a, b \le H$ and for which

$$ax^2 + by^2 = z^2$$

has a non-trivial solution.

Before applying Turán's sieve, let us show a few auxiliary results.

Lemma 4.5.1 *Let $p \equiv 1 \pmod 4$. The value of $(a, b)_p$ depends only on the reduced residue classes of a and b modulo p^2 if neither of them is divisible by p^2.*

Proof We write $a = p^2 t + a_1, b = p^2 s + b_1$, with $1 \le a_1, b_1 \le p^2 - 1$. If a_1, b_1 are both coprime to p, then $(a, b)_p = (a_1, b_1)_p = 1$. If $p | a_1$ and b_1 is coprime to p, then

$$(a, b)_p = \left(\frac{b_1}{p} \right).$$

The case $p | b_1$ and a_1 coprime to p is similar. If $p | a_1$ and $p | b_1$ then $(a, b)_p = (a_1, b_1)_p$. This completes the proof. $\square$

Lemma 4.5.2 *Let $p \equiv 1 \pmod 4$. Then*

$$\# \mathcal{A}_p = \frac{H^2}{p} + O \left(\frac{H^2}{p^2} + p^4 + Hp^2 \right).$$

Proof We first remove from $\mathcal{A}_p$ those elements (a, b) with at least one component divisible by p^2. The number of such pairs is clearly $O(H^2/p^2)$. The remaining elements are enumerated by the sum

$$\frac{1}{2} \sum_{\substack{1 \le a, b \le H \\ p^2 \nmid a, b}} \left(1 - (a, b)_p \right).$$

By the previous lemma, the value of $(a, b)_p$ depends only on the reduced classes of $a, b \pmod{p^2}$. Thus

$$\sum_{\substack{1 \le a, b \le H \\ p^2 \nmid a, b}} (a, b)_p = \sum_{1 \le a_1, b_1 \le p^2 - 1} (a_1, b_1)_p \left(\frac{H}{p^2} + O(1) \right)^2$$

$$= \frac{H^2}{p^4} \sum_{1 \le a_1, b_1 \le p^2 - 1} (a_1, b_1)_p + O \left(p^4 + Hp^2 \right).$$

The summation on the right-hand side is easily determined. If a_1, b_1 are both coprime to p, we have $(a_1, b_1)_p = 1$ and the contribution to the sum from such terms is

$$[p(p-1)]^2.$$

If $p|a_1$ and b_1 is coprime to p, then

$$(a_1, b_1)_p = \left(\frac{b_1}{p}\right)$$

and the contribution to the sum from such terms is zero. The same is the case if $p|b_1$ and a_1 is coprime to p. If $p|a_1$, $p|b_1$ then $a_1 = b_1 = p$ and we get an error of $O(H^2/p^4)$.

Hence the final result is that

$$\#\mathcal{A}_p = \frac{H^2}{p} + O\left(\frac{H^2}{p^2} + p^4 + Hp^2\right).$$

$\square$

Lemma 4.5.3 *Let $p, q \equiv 1 \pmod 4$ be distinct primes. Then*

$$\#\mathcal{A}_p \cap \mathcal{A}_q = \frac{H^2}{pq} + O\left(\frac{H^2}{p^2} + \frac{H^2}{q^2} + p^4 q^4 + Hp^2 q^2\right).$$

Proof As before, we remove from $\mathcal{A}_p$ and $\mathcal{A}_q$ any pair (a, b) with at least one of a, b divisible by p^2 or q^2. The remaining elements are enumerated by

$$\frac{1}{4} \sum_{\substack{1 \le a, b \le H \\ p^2 \nmid a, b \\ q^2 \nmid a, b}} \left(1 - (a, b)_p\right)\left(1 - (a, b)_q\right).$$

Expanding the product in the summand leads to the consideration of four sums, three of which can be treated by the previous lemma. Thus we need only consider

$$\sum_{\substack{1 \le a, b \le H \\ p^2, q^2 \nmid a, b}} (a, b)_p (a, b)_q.$$

By Lemma 4.5.1, each of the summands depends only on the residue class of $a, b \pmod{p^2 q^2}$. As before, we partition the sum according to the residue classes modulo pq and find that it equals

$$H^2 \left(\frac{p^2 - p}{p^2}\right)\left(\frac{q^2 - q}{q^2}\right) + O\left((pq)^4 + Hp^2 q^2\right).$$

By putting everything together we find

$$\#\mathcal{A}_p \cap \mathcal{A}_q = \frac{H^2}{pq} + O\left(\frac{H^2}{p^2} + \frac{H^2}{q^2} + (pq)^4 + Hp^2q^2\right),$$

as desired. □

We can now apply the Turán sieve to prove:

Theorem 4.5.4 *The number of pairs of integers (a,b) with $1 \leq a, b \leq H$ and for which $ax^2 + by^2 = z^2$ has a non-trivial rational point is*

$$\ll \frac{H^2}{\log\log H}.$$

Proof By the Turán sieve we find that the number in question is

$$\ll \frac{H^2}{\log\log z} + z^{10} + Hz^6,$$

where $\mathcal{P}$ is chosen to be the set of primes $p \equiv 1 \pmod 4$ with $p \leq z$. Choosing $z := H^{(1/6)-\varepsilon}$ for any $\varepsilon > 0$ gives us the required result. □

Remark 4.5.5 It can be seen, by invoking the large sieve, that this estimate can be refined to

$$\ll \frac{H}{(\log H)^\delta}$$

for some $\delta > 0$. This is proven in [65].

4.6 Exercises

1. With N_n as defined in Theorem 4.2.1, show that

$$N_n = \frac{p^n}{n} + O\left(p^{n/2}d(n)\right),$$

   where $d(n)$ denotes the number of divisors of n.
2. Show that $x^4 + 6x^2 + 1$ is irreducible in $\mathbb{Z}[x]$, but reducible modulo p for every prime p. [Hint: consider $(x^2 + bx \pm 1)(x^2 - bx \pm 1) \pmod p$ for suitable b.] (One can prove that an irreducible polynomial $f(x) \in \mathbb{Z}[x]$ of degree n is reducible modulo p for every prime p if and only if the Galois group of f has no cycles of order n. This is a consequence of the Chebotarev density theorem.)

3. Prove that the number of irreducible polynomials

$$x^n + a_{n-1}x^{n-1} + \cdots + a_1 x + a_0, \quad 0 \le a_i < H, \ a_i \in \mathbb{Z},$$

   that are reducible modulo p for every prime p is $O\left(H^{n-1/3}(\log H)^{2/3}\right)$.

4. Let $\mathcal{A}$ be the set of natural numbers $\le x$ and let $\mathcal{P}$ be the set of primes $p < z$. Use Theorem 4.1.1 to show that

$$\pi(x) \ll \frac{x}{\log \log x}.$$

   Note that only a lower bound for $U(z)$ is required and hence the relevance of Remark 1.4.5 made at the end of Chapter 1, where Chebycheff's theorem was not used.

5. Prove that the number of reducible polynomials

$$x^n + a_{n-1}x^{n-1} + \cdots + a_1 x + a_0, \quad 0 \le a_i \le H, \ a_i \in \mathbb{Z},$$

   is $\gg H^{n-1}$.

6. Show that the number of reducible polynomials $x^2 + ax + b$ with $a, b \in \mathbb{Z}$ and $0 \le a, b \le H$ is $O(H \log H)$. Can this estimate be sharpened?

7. If $k = \mathbb{R}$, show that the Hilbert symbol (a, b) is 1 if a or b is > 0 and is -1 if a and b are < 0.

8. Modify the proof of the main result of Section 4.5 to deduce that the number of integers a, b with $|a|, |b| \le H$ and for which $ax^2 + by^2 = z^2$ has a non-trivial rational solution is

$$O\left(\frac{H^2}{\log \log H}\right).$$

9. Prove that there is a constant $c > 0$ such that

$$\sum_{n \le x} (\nu(n) - \log \log x)^2 = x \log \log x + cx + O\left(\frac{x \log \log x}{\log x}\right).$$

10. Let $d(n)$ denote the number of divisors of n. Show that $\log d(n)$ has normal order $(\log 2) \log \log n$. [Hint: let $\Omega(n)$ be the number of prime factors of n, counted with multiplicity, then $2^{\nu(n)} \le d(n) \le 2^{\Omega(n)}$.]

11. Prove the inequality

$$\sum_{n \le x} \left| \sum_{\substack{p \mid n \\ p > x^{1/2}}} f(p) - \sum_{x^{1/2} < p \le x} \frac{f(p)}{p} \right|^2 \le 2x \sum_{x^{1/2} < p \le x} \frac{|f(p)|^2}{p}$$

   for all complex numbers $f(p)$ and $x \ge 1$. [Hint: first treat the case $f(p)$ real and non-negative and then apply the methods of Exercises 17 and 18 of Chapter 3 to deal with the general case.]

12. Let $\Psi(x, y)$ denote the number of $n \leq x$ all of whose prime factors are $\leq y$. Let $\varepsilon > 0$ and $y = x^\varepsilon$. By considering numbers of the form

$$n = m p_1 \dots p_k \leq x$$

with

$$x^{\varepsilon - \varepsilon^2} < p_i < x^\varepsilon, \quad 1 \leq i \leq k = [1/\varepsilon],$$

show that

$$\Psi(x, y) \gg x,$$

where the implied constant depends on ε.

13. Let $\varepsilon > 0$ be fixed. By applying the method of proof of Theorem 4.1.1, show that the number of primes $p \leq x$ such that the smallest quadratic non-residue modulo p is greater than x^ε is bounded by a constant depending on ε. [Hint: let $\mathcal{A}$ be the set of numbers $n \leq x$ and let $\mathcal{P}$ be the set of primes $p \leq x^{1/4}$ for which the least quadratic non-residue is $> x^\varepsilon$. Let $\mathcal{A}_p$ be the set of elements of $\mathcal{A}$ that are quadratic non-residues modulo p. Observe that $S(\mathcal{A}, \mathcal{P}, x^{1/4}) \geq \Psi(x, x^\varepsilon)$.] (This result was first obtained by U. V. Linnik (1915–72) as an application of his large sieve inequality.)

14. Let f be a non-negative valued function and let $\mathcal{A}, \mathcal{P}, z$ be as in Section 4.1. Assume that (4.1) and (4.2) hold. Define, for $a \in \mathcal{A}$,

$$N_f(a) := \sum_{p:a\in\mathcal{A}_p} f(p),$$

$$U_f(z) := \sum_{p|P(z)} \delta_p f(p).$$

Show that

$$\sum_{a\in\mathcal{A}} (N_f(a) - U_f(z))^2 = X \sum_{p|P(z)} f(p)^2 \delta_p (1 - \delta_p)$$

$$+ \sum_{p,q|P(z)} f(p)f(q)R_{p,q} - 2U_f(z) \sum_{p|P(z)} f(p)R_p.$$

15. With notation as in the previous exercise and Theorem 4.1.1, show that

$$S(\mathcal{A}, \mathcal{P}, z) \leq \frac{XV_f(z)}{U_f(z)^2} + \frac{2}{U_f(z)} \sum_{p|P(z)} f(p)|R_p|$$

$$+ \frac{1}{U_f(z)^2} \sum_{p,q|P(z)} f(p)f(q)|R_{p,q}|,$$

where

$$V_f(z) := \sum_{p|P(z)} \delta_p f(p)^2.$$

16. With U_f and V_f as in the previous exercise, show that

$$U_f(z)^2 \leq U(z)V_f(z).$$

Infer that the optimal estimate for $S(\mathcal{A}, \mathcal{P}, z)$ implied by Exercise 15 is the one given by Theorem 4.1.1 (namely, the choice $f \equiv 1$).

17. Let $\mathcal{A}$ and $\mathcal{P}$ be as in Section 4.1. For each $a \in \mathcal{A}$, let $w(a)$ be non-negative real number (usually called a 'weight') and set

$$X := \sum_{a \in \mathcal{A}} w(a).$$

Suppose that, for $p \in \mathcal{P}$,

$$\sum_{a \in \mathcal{A}_p} w(a) = \delta_p X + R_p$$

and, for distinct primes $p, q \in \mathcal{P}$,

$$\sum_{a \in \mathcal{A}_{pq}} w(a) = \delta_p \delta_q X + R_{p,q}.$$

Let $N(a)$ be the number of $p \in \mathcal{P}$ for which $a \in \mathcal{A}_p$ and let

$$U(z) := \sum_{p|P(z)} \delta_p.$$

Prove that

$$\sum_{a \in \mathcal{A}} w(a)\, (N(a) - U(z))^2 = X \sum_{p|P(z)} \delta_p (1 - \delta_p) + \sum_{p,q|P(z)} R_{p,q} - 2U(z) \sum_{p|P(z)} R_p.$$

18. Let $y = \log x$ and for each natural number $n \leq x$ define $w(n)$ to be 1 if n has no prime factor $\leq y$ and zero otherwise. Show that

$$\sum_{n \leq x} w(n)\, (\nu(n) - \log\log x)^2 = xV(y) \log\log x + O(xV(y)),$$

where

$$V(y) := \prod_{p \leq y} \left(1 - \frac{1}{p} \right).$$

[Hint: let $P(y)$ denote the product of the primes $\leq y$. Observe that

$$w(n) = \sum_{d|(n,P(y))} \mu(d).$$

Now apply the previous exercise.]

19. Deduce from the previous exercise that the number of primes $p \leq x$ is

$$O\left(\frac{x}{(\log \log x)^2}\right).$$

Compare this with the result obtained in Exercise 4 above.

5

The sieve of Eratosthenes

Eratosthenes (276–194 BCE) was the director of the famous library of Alexandria. The sieve of Eratosthenes was first described in the work of Nicomedes (280–210 BCE), entitled *Introduction to Arithmetic*.

The form of the sieve of Eratosthenes we describe in Section 5.1 below is due to A. M. Legendre (1752–1833) and was written down by him in 1808 in the second edition of his book *Théorie des Nombres*.

In this chapter, we treat this material from a modern perspective and show that when it is combined with 'Rankin's trick', the sieve of Eratosthenes becomes as powerful as Brun's pure sieve, the latter being more cumbersome and awkward to derive.

5.1 The sieve of Eratosthenes

We will use the fundamental property of the Möbius function given in Lemma 1.2.1 to study the number

$$\Phi(x, z) := \# \{ n \leq x : n \text{ is not divisible by any prime} < z \},$$

where x, z are positive real numbers. This will be used as a motivating example that will signal a more formal development of the sieve of Eratosthenes, undertaken in a later section.

If we let

$$P_z := \prod_{p<z} p,$$

then

$$\Phi(x, z) = \sum_{n \le x} \sum_{d \mid (n, P_z)} \mu(d) = \sum_{d \mid P_z} \mu(d) \left[\frac{x}{d} \right]$$

$$= x \sum_{d \mid P_z} \frac{\mu(d)}{d} + O(2^z) = x \prod_{p < z} \left(1 - \frac{1}{p} \right) + O(2^z).$$

We can use this to deduce an elementary estimate for $\pi(x)$, as follows.
First note that

$$\pi(x) = (\pi(x) - \pi(z)) + \pi(z)$$

$$\le \Phi(x, z) + \pi(z)$$

$$\le \Phi(x, z) + z. \tag{5.1}$$

Moreover, the inequality $1 - x \le e^{-x}$, valid for positive x, implies that

$$\prod_{p < z} \left(1 - \frac{1}{p} \right) \le \exp \left(-\sum_{p < z} \frac{1}{p} \right).$$

In Chapter 1 we established in an elementary fashion (see Remark 1.4.5) that

$$\sum_{p < z} \frac{1}{p} \ge \log \log z + O(1).$$

Thus we obtain an upper bound for

$$\prod_{p < z} \left(1 - \frac{1}{p} \right),$$

and consequently for $\Phi(x, z)$. Choosing

$$z := c \log x$$

for some small positive constant c and using (5.1), we find:

Proposition 5.1.1

$$\pi(x) \ll \frac{x}{\log \log x}.$$

In the next two sections we will investigate ways of improving the above analysis. As we will see in Section 5.4, this analysis has an interesting application to the study of twin primes.

5.2 Mertens' theorem

We begin by deriving an asymptotic formula for

$$V(z) := \prod_{p < z} \left(1 - \frac{1}{p} \right).$$

In Section 5.1 we only obtained an upper bound for this product. It is easy to derive the asymptotic formula for it, and we do it below. However, to determine the constants explicitly in the formula will involve an application of some properties of the Riemann zeta function $\zeta(s)$. Most notably, we will need to use the fact that $\zeta(s)$ can be extended analytically to the region $\mathrm{Re}(s) > 0$, apart from a simple pole at $s = 1$ with residue 1.

We write

$$-\log V(z) = \sum_{p < z} \frac{1}{p} + \sum_{\substack{k \geq 2 \\ p < z}} \frac{1}{k p^k},$$

and since

$$\sum_{\substack{k \geq 2 \\ p < z}} \frac{1}{k p^k} \leq \sum_{p < z} \sum_{k \geq 2} \frac{1}{p^k} = \sum_{p < z} \frac{1}{p(p-1)},$$

we may write

$$\sum_{\substack{k \geq 2 \\ p < z}} \frac{1}{k p^k} = c_0 + O \left(\frac{1}{z} \right)$$

for some positive constant c_0. We deduce that

$$-\log V(z) = \sum_{p < z} \frac{1}{p} + c_0 + O \left(\frac{1}{z} \right),$$

where, actually,

$$c_0 = -\sum_{p} \left\{ \log \left(1 - \frac{1}{p} \right) + \frac{1}{p} \right\}.$$

Recall that by Theorem 1.4.3 of Chapter 1 we have

$$R(z) := \sum_{p < z} \frac{\log p}{p} = \log z + O(1),$$

so that, by partial summation,

$$\sum_{p<z} \frac{1}{p} = \frac{R(z)}{\log z} + \int_2^z \frac{R(t)dt}{t \log^2 t}$$

$$= \log \log z + c_1 + O\left(\frac{1}{\log z}\right)$$

for some positive constant c_1. Thus

$$-\log V(z) = \log \log z + c_0 + c_1 + O\left(\frac{1}{\log z}\right),$$

and so

$$\prod_{p<z}\left(1 - \frac{1}{p}\right) = \frac{e^{-(c_0+c_1)}}{\log z}\left(1 + O\left(\frac{1}{\log z}\right)\right)$$

as $z \to \infty$.

By a careful analysis, Mertens showed that $c_0 + c_1 = \gamma$, where γ is Euler's constant defined in Exercise 7 of Chapter 1. This is derived as follows. For $\sigma > 0$ consider

$$\zeta(1+\sigma) = \sum_{n=1}^{\infty} \frac{1}{n^{1+\sigma}}$$

and recall that we have the Euler product

$$\zeta(1+\sigma) = \prod_{p}\left(1 - \frac{1}{p^{1+\sigma}}\right)^{-1}.$$

Thus

$$f(\sigma) := \log \zeta(1+\sigma) - \sum_{p} \frac{1}{p^{1+\sigma}}$$

$$= -\sum_{p}\left\{\log\left(1 - \frac{1}{p^{1+\sigma}}\right) + \frac{1}{p^{1+\sigma}}\right\},$$

so that

$$c_0 = \lim_{\sigma \to 0} f(\sigma).$$

By Exercise 2 we know that, for $\sigma > 0$,

$$\log \zeta(1+\sigma) = \log \frac{1}{\sigma} + O(\sigma)$$

$$= -\log(1 - e^{-\sigma}) + O(\sigma)$$

$$= \sum_{n=1}^{\infty} e^{-\sigma n} n^{-1} + O(\sigma).$$

Put

$$H(t) := \sum_{n \le t} \frac{1}{n}$$

and

$$P(t) := \sum_{p \le t} \frac{1}{p}.$$

Then, by partial summation,

$$\sum_p \frac{1}{p^{1+\sigma}} = \sigma \int_1^\infty \frac{P(u)}{u^{1+\sigma}} du = \sigma \int_0^\infty P(e^t) e^{-\sigma t} dt.$$

Similarly,

$$\log \zeta(1+\sigma) = \sigma \int_0^\infty e^{-\sigma t} H(t) dt + O(\sigma).$$

Hence

$$f(\sigma) = \sigma \int_0^\infty e^{-\sigma t} \left(H(t) - P(e^t) \right) dt + O(\sigma).$$

Since

$$H(t) = \log t + \gamma + O\left(\frac{1}{t}\right)$$

and

$$P(e^t) = \log t + c_1 + O\left(\frac{1}{t}\right),$$

we deduce

$$f(\sigma) = \sigma \int_0^\infty e^{-\sigma t} \left(\gamma - c_1 + O\left(\frac{1}{t+1}\right) \right) dt + O(\sigma)$$

$$= \gamma - c_1 + O(\sigma).$$

Letting $\sigma \to 0$ gives

$$f(0) = c_0 = \gamma - c_1.$$

This proves:

Theorem 5.2.1

$$V(z) = \prod_{p<z} \left(1 - \frac{1}{p} \right) = \frac{e^{-\gamma}}{\log z} \left(1 + O\left(\frac{1}{\log z} \right) \right)$$

as $z \to \infty$.

5.3 Rankin's trick and the function $\Psi(x,z)$

If we consider

$$\Psi(x,z) := \#\{n \le x: \text{ if } p|n, \text{ then } p < z\},$$

we find that in the sieve of Eratosthenes discussed in Section 5.1 we can make a finer analysis:

$$\Phi(x,z) = \sum_{\substack{d|P_z \\ d\le x}} \mu(d)\left[\frac{x}{d}\right] = x\sum_{\substack{d|P_z \\ d\le x}} \frac{\mu(d)}{d} + O(\Psi(x,z)). \tag{5.2}$$

We now use a **trick of Rankin** to estimate $\Psi(x,z)$. For any $\delta > 0$, we have

$$\Psi(x,z) = \sum_{\substack{n\le x \\ p|n\Rightarrow p<z}} 1 \le \sum_{\substack{n\le x \\ p|n\Rightarrow p<z}} \left(\frac{x}{n}\right)^\delta \le x^\delta \prod_{p<z}\left(1-\frac{1}{p^\delta}\right)^{-1}.$$

We observe that

$$\prod_{p<z}\left(1-\frac{1}{p^\delta}\right)^{-1} \ll \prod_{p<z}\left(1+\frac{1}{p^\delta}\right)\left(1-\frac{1}{p^{2\delta}}\right)^{-1} \ll \prod_{p<z}\left(1+\frac{1}{p^\delta}\right),$$

since the product

$$\prod_p\left(1-\frac{1}{p^{2\delta}}\right)^{-1}$$

converges for $\delta > 1/2$. Therefore

$$\Psi(x,z) \ll x^\delta \prod_{p<z}\left(1+\frac{1}{p^\delta}\right).$$

By applying the elementary inequality $1+x \le e^x$ we obtain

$$\Psi(x,z) \ll \exp\left(\delta\log x + \sum_{p<z}\frac{1}{p^\delta}\right).$$

Now we choose $\delta := 1-\eta$ with $\eta\to0$ as $z\to\infty$. By writing

$$p^{-\delta} = p^{-1}p^\eta = p^{-1}e^{\eta\log p}$$

and using the inequality $e^x \le 1+xe^x$, we deduce

$$\sum_{p<z}\frac{1}{p^\delta} \le \sum_{p<z}\frac{1}{p}(1+(\eta\log p)z^\eta),$$

since $p < z$. Thus, choosing

$$\eta := \frac{1}{\log z}$$

gives:

Theorem 5.3.1

$$\Psi(x, z) \ll x(\log z) \exp\left(-\frac{\log x}{\log z}\right)$$

as $x \to \infty$.

We apply the above result to deduce:

Theorem 5.3.2

$$\sum_{\substack{d \mid P_z \\ d \leq x}} \frac{\mu(d)}{d} = \prod_{p < z}\left(1 - \frac{1}{p}\right) + O\left((\log z)^2 \exp\left(-\frac{\log x}{\log z}\right)\right).$$

Proof   We have

$$\sum_{\substack{d \mid P_z \\ d \leq x}} \frac{\mu(d)}{d} = \prod_{p < z}\left(1 - \frac{1}{p}\right) - \sum_{\substack{d \mid P_z \\ d > x}} \frac{\mu(d)}{d}.$$

By Exercise 3, the last sum is dominated by

$$\sum_{\substack{d \mid P_z \\ d > x}} \frac{1}{d} \leq -\frac{\Psi(x, z)}{x} + \int_x^\infty \frac{\Psi(t, z)}{t^2} dt.$$

The integral is bounded by

$$(\log z) \int_x^\infty \exp\left(-\frac{\log t}{\log z}\right)\frac{dt}{t} = (\log z)\int_x^\infty \frac{dt}{t^{1+1/\log z}} \leq (\log z)^2 \exp\left(-\frac{\log x}{\log z}\right),$$

which completes the proof of the theorem. $\square$

Going back to (5.2) and combining the above theorems, we see that we have proven:

Theorem 5.3.3

$$\Phi(x, z) = x\prod_{p < z}\left(1 - \frac{1}{p}\right) + O\left(x(\log z)^2 \exp\left(-\frac{\log x}{\log z}\right)\right)$$

as $x, z \to \infty$.

Corollary 5.3.4

$$\pi(x) \ll \frac{x}{\log x}(\log \log x).$$

Proof In the theorem choose $\log z = \varepsilon \log x / \log \log x$ for some sufficiently small ε and note that

$$\prod_{p<z}\left(1-\frac{1}{p}\right) \leq \exp\left(-\sum_{p<z}\frac{1}{p}\right) \ll \frac{1}{\log z}.$$

By applying (5.1) we get the result. We emphasize that the last estimate above does not use Chebycheff's theorem. $\square$

5.4 The general sieve of Eratosthenes and applications

We are ready to formalize our discussions. Let $\mathcal{A}$ be any set of natural numbers $\leq x$ and let $\mathcal{P}$ be a set of primes. To each prime $p \in \mathcal{P}$, let there be $\omega(p)$ distinguished residue classes modulo p. Let $\mathcal{A}_p$ denote the set of elements of $\mathcal{A}$ belonging to at least one of these distinguished residue classes modulo p, let $\mathcal{A}_1 := \mathcal{A}$ and for any squarefree integer d composed of primes in $\mathcal{P}$ define

$$\mathcal{A}_d := \cap_{p|d}\mathcal{A}_p$$

and

$$\omega(d) := \prod_{p|d}\omega(p).$$

Let z be a positive real number and set

$$P(z) := \prod_{\substack{p\in\mathcal{P}\\ p<z}} p.$$

As usual, we denote by $S(\mathcal{A}, \mathcal{P}, z)$ the number of elements of

$$\mathcal{A}\setminus \cup_{p|P(z)}\mathcal{A}_p.$$

We suppose that there is an X such that

$$\#\mathcal{A}_d = \frac{\omega(d)}{d}X + R_d \tag{5.3}$$

for some R_d.

Theorem 5.4.1 *(The sieve of Eratosthenes)*
In the above setting, suppose that the following conditions are satisfied:

1. $|R_d| = O(\omega(d))$;
2. *for some $\kappa \geq 0$,*

$$\sum_{p|P(z)} \frac{\omega(p)\log p}{p} \leq \kappa \log z + O(1);$$

3. *for some positive real number y, $\#\mathcal{A}_d = 0$ for every $d > y$.*

Then

$$S(\mathcal{A}, \mathcal{P}, z) = XW(z) + O\left(\left(X + \frac{y}{\log z}\right)(\log z)^{\kappa+1}\exp\left(-\frac{\log y}{\log z}\right)\right),$$

where

$$W(z) := \prod_{\substack{p\in\mathcal{P}\\ p<z}}\left(1 - \frac{\omega(p)}{p}\right).$$

To prove this, we will need the following lemmas.

Lemma 5.4.2 *With the setting and hypotheses of Theorem 5.4.1, let*

$$F(t, z) := \sum_{\substack{d\leq t\\ d|P(z)}} \omega(d).$$

Then

$$F(t, z) = O\left(t(\log z)^{\kappa}\exp\left(-\frac{\log t}{\log z}\right)\right).$$

Proof As before, we use Rankin's trick. For any $\delta > 0$,

$$F(t, z) \leq \sum_{d|P(z)} \omega(d)\left(\frac{t}{d}\right)^{\delta}.$$

Since ω is multiplicative, we deduce that

$$F(t, z) \leq \exp\left(\delta\log t + \sum_{p|P(z)} \frac{\omega(p)}{p^{\delta}}\right)$$

by applying the elementary inequality $1 + x \leq e^x$. Setting $\delta := 1 - \eta$ and using the inequality $e^x \leq 1 + xe^x$, we find

$$F(t, z) \leq t\exp\left(-\eta\log t + \sum_{p|P(z)} \frac{\omega(p)}{p} + \eta z^{\eta}\sum_{p|P(z)} \frac{\omega(p)\log p}{p}\right).$$

By the second hypothesis of Theorem 5.4.1 and partial summation we find that

$$\sum_{p|P(z)} \frac{\omega(p)}{p} \leq \kappa \log \log z + O(1).$$

Thus

$$F(t, z) \ll t \exp\left(-\eta \log t + \kappa \log \log z + \kappa \eta (\log z)^{\eta}\right).$$

Choosing $\eta := 1/\log z$, as before, gives the result. $\square$

Lemma 5.4.3 *With the setting and hypotheses of Theorem 5.4.1,*

$$\sum_{\substack{d|P(z) \\ d>y}} \frac{\omega(d)}{d} = O\left((\log z)^{\kappa+1} \exp\left(-\frac{\log y}{\log z}\right)\right).$$

Proof Clearly, by partial summation we have

$$\sum_{\substack{d|P(z) \\ d>y}} \frac{\omega(d)}{d} \ll \int_y^\infty \frac{F(t, z)}{t^2} dt.$$

The result now follows from Lemma 5.4.2. $\square$

Proof of Theorem 5.4.1 By the inclusion–exclusion principle and the first and third hypotheses,

$$S(\mathcal{A}, \mathcal{P}, z) = \sum_{\substack{d|P(z) \\ d\leq y}} \mu(d) \# \mathcal{A}_d$$

$$= \sum_{\substack{d|P(z) \\ d\leq y}} \mu(d) \frac{X\omega(d)}{d} + O(F(y, z)).$$

Then by Lemmas 5.4.2 and 5.4.3 we find easily that

$$S(\mathcal{A}, \mathcal{P}, z) = XW(z) + O\left(\left(X + \frac{y}{\log z}\right)(\log z)^{\kappa+1} \exp\left(-\frac{\log y}{\log z}\right)\right).$$

$\square$

We recall that a prime number p such that $p+2$ is also a prime is called a **twin prime** and that the famous **twin prime conjecture**, still unresolved, predicts that there are infinitely many twin primes. One can say that this conjecture has spawned extensive research in sieve theory. In fact, the birth of modern sieve theory can be traced to a 1919 paper of Viggo Brun (see [27] for example), where he proved that

$$\sum_{\substack{p \\ p+2\,\text{prime}}} \frac{1}{p} < \infty.$$

Before Brun's paper, no one knew how to tackle such questions. Brun deduced this result from an estimate he obtained by a sophisticated sieve method, now called Brun's sieve, which we will discuss in the next chapter. The point of theoretical interest we now make is that Brun's bound can be derived from the elementary sieve of Eratosthenes discussed here. The upper bound derived below is not best possible, but will suffice to deduce the convergence of the above series.

Theorem 5.4.4 *The number of primes $p \leq x$ such that $p+2$ is prime is*

$$\ll \frac{x(\log \log x)^2}{\log^2 x}.$$

Proof Let $\mathcal{A}$ be the set of natural numbers $n \leq x$ and let $\mathcal{P}$ be the set of all primes. Let $z = z(x)$ be a positive real number, to be chosen soon. For each prime $p < z$ we distinguish the residue classes 0 and -2 modulo p. Since $\mathcal{A}_p$ (the set of $n \leq x$ belonging to at least one of these residue classes) is empty for $p > x+2$, we apply Theorem 5.4.1 with $\kappa = 2$ to deduce that

$$S(\mathcal{A}, \mathcal{P}, z) = xW(z) + O\left(x(\log z)^3 \exp\left(-\frac{\log x}{\log z}\right)\right),$$

where

$$W(z) := \prod_{p<z}\left(1 - \frac{2}{p}\right).$$

Now

$$W(z) = \prod_{p<z}\left(1 - \frac{2}{p}\right) \leq \exp\left(-\sum_{p<z}\frac{2}{p}\right) \ll (\log z)^{-2}.$$

We choose z such that

$$\log z = \log x / A \log \log x$$

for some large positive constant A and deduce

$$S(\mathcal{A}, \mathcal{P}, z) \ll \frac{x(\log \log x)^2}{\log^2 x}.$$

Clearly, the number of twin primes cannot exceed

$$\pi(z) + S(\mathcal{A}, \mathcal{P}, z) \leq z + S(\mathcal{A}, \mathcal{P}, z),$$

with z as above. The result is now immediate. $\square$

Corollary 5.4.5 *(Brun's theorem)*
The sum

$$\sum_{\substack{p \\ p+2 \text{ prime}}} \frac{1}{p}$$

converges.

Proof By partial summation and Theorem 5.4.4, the sum is bounded by

$$\ll \int_2^\infty \frac{(\log \log t)^2 \mathrm{d}t}{t \log^2 t},$$

which is finite. $\square$

Remark 5.4.6 The elementary reasoning used to derive the general sieve of Eratosthenes actually yields a better result than stated; we have chosen the above parameters to keep the exposition simple. More precisely, if

$$2\kappa z \log z > \log x,$$

then

$$S(\mathcal{A}, \mathcal{P}, z) = XW(z) + O\left(\left(X + \frac{y}{\log z}\right)\right.$$

$$\left. \times (\log z)^{\kappa+1} \exp\left(-\frac{\log y}{2 \log z} \log\left(\frac{\log y}{2\kappa \log z}\right)\right)\right).$$

This is obtained by setting

$$\eta := \frac{1}{\log z} \log\left(\frac{\log x}{2\kappa \log z}\right) < 1$$

in the derivation of Theorems 5.4.1 and 5.4.2.

5.5 Exercises

1. Show that

$$\sum_{d^2 \mid n} \mu(d) = \begin{cases} 1 & \text{if } n \text{ is squarefree} \\ 0 & \text{otherwise.} \end{cases}$$

If $Q(x)$ denotes the number of squarefree numbers $\leq x$, deduce that

$$Q(x) = \frac{6}{\pi^2}x + O(\sqrt{x}).$$

2. By using partial summation, show that

$$\zeta(1+\sigma) = \frac{1}{\sigma} + O(\sigma)$$

for $\sigma > 0$.

3. Let $C(x) = \sum_{n \leq x} c_n$ and let $f(t)$ be a differentiable function with continuous derivative. Suppose that

$$\lim_{Y \to \infty} C(Y)f(Y) = 0$$

and

$$\int_1^\infty C(t)f'(t)dt < \infty.$$

Prove that

$$\sum_{n > x} c_n f(n) = -C(x)f(x) - \int_x^\infty C(t)f'(t)dt.$$

4. Let $C(x)$ be as in the previous exercise. Suppose that there is a $\sigma > 0$ so that $C(x)/x^\sigma \to 0$ as $x \to \infty$. Show that

$$\sum_{n=1}^\infty \frac{c_n}{n^s} = s\int_1^\infty \frac{C(t)}{t^{s+1}}dt$$

for $\mathrm{Re}(s) > \sigma$.

5. Prove that

$$\sum_{p \leq x/2} \frac{1}{p \log \frac{x}{p}} = O\left(\frac{\log \log x}{\log x}\right)$$

by subdividing the interval $[1, x/2]$ into subintervals of the form $I_j = [e^j, e^{j+1}]$. (This technique is referred to as the **method of dyadic subdivision**.)

6. Let $\pi_k(x)$ denote the number of $n \leq x$ with k prime factors (not necessarily distinct). Using the sieve of Eratosthenes, show that

$$\pi_k(x) \leq \frac{x(A \log \log x + B)^k}{k! \log x}$$

for some positive constants A and B.

7. Using Chebycheff's estimate for $\pi(x)$, show that if $0 \le \delta < 1$, then

$$\sum_{p \le z} \frac{1}{p^\delta} \ll \frac{z^{1-\delta}}{1-\delta},$$

where the implied constant is absolute. Using this refined estimate in the proof of Theorem 5.3.1, deduce that

$$\Psi(x, z) \ll x \exp\left(-\frac{\log x}{\log z}\right).$$

8. Using the method of the previous exercise, show that for some constant $c > 0$,

$$\Phi(x, z) \ll x \exp\left(-\eta \log x + \frac{cz^\eta}{\eta \log z}\right),$$

for any $0 < \eta < 1$. Choosing

$$\eta := \frac{1}{\log z} \log\left(\frac{\log x}{\log z}\right),$$

which is less than 1 if $\log x < z \log z$, deduce that, for some $c_1 > 0$,

$$\Phi(x, z) \ll x \exp\left(-\frac{c_1 \log x}{\log z} \log\left(\frac{\log x}{\log z}\right)\right).$$

Apply this result to deduce that Corollary 5.3.4 can be improved to

$$\pi(x) \ll \frac{x \log \log x}{(\log x) \log \log \log x}.$$

(Compare this with the technique of Exercises 23–25 below.)

9. Let $(a, k) = 1$ and let $\pi(x; k, a)$ denote the number of primes $p \le x$ $p \equiv a \pmod{k}$. Let $A > 0$. Show that

$$\pi(x; k, a) \ll \frac{x \log \log x}{\phi(k) \log x}$$

uniformly for $k \le (\log x)^A$, where the implied constant depends only on A.

10. Let $f(n)$ be a **completely multiplicative function** (that is, $f(mn) = f(m)f(n)$ for any m, n) satisfying $0 \le f(p) < A$. Let x, z be positive real numbers. Show that

$$\sum_{\substack{n \le x \\ p \mid n \Rightarrow p < z}} f(n) \ll x(\log z)^A \exp\left(-\frac{\log x}{\log z}\right).$$

11. Let $\mathrm{rad}(n)$ denote the product of the prime divisors of n. We set $\mathrm{rad}(1) = 1$. Show that the Dirichlet series

$$\sum_{n=1}^{\infty} \frac{1}{\mathrm{rad}(n)n^s}$$

converges for $\mathrm{Re}(s) > 0$. Using this fact, deduce that

$$N(x, y) := \#\{n \le x : \mathrm{rad}(n) \le y\}$$

is $O(yx^{\varepsilon})$ for any $\varepsilon > 0$.

12. With notation as in the previous exercise, show that for some constant $c > 0$,

$$N(x, y) \ll y(\log y)\exp(c\sqrt{\log(x/y)})$$

for $2 \le y \le x$. [Hint: for any natural number n enumerated by $N(x, y)$, notice that

$$1 \le \left(\frac{x}{n}\right)^{\delta}\left(\frac{y}{x\,\mathrm{rad}(n)}\right)^{1-\delta}$$

for any $0 \le \delta \le 1$.]

13. Let $a_1, a_2, \ldots, a_k$ be positive real numbers. Let $w > 0$ and n be such that $w \ge na_k$. By comparing

$$\sum_{j=0}^{n}(w - ja_k)^{k-1} \quad \text{and} \quad \int_0^n (w - ta_k)^{k-1}\,dt,$$

deduce that

$$\frac{w^k}{ka_k} < \sum_{j=0}^{n}(w - ja_k)^{k-1} < \frac{(w + a_k)^k}{ka_k}.$$

14. Let $a_1, \ldots, a_k$ be positive real numbers. Let $N(a_1, \ldots, a_k; z)$ be the number of k-tuples $(n_1, \ldots, n_k)$ such that $n_i \ge 0$ and

$$n_1 a_1 + \cdots + n_k a_k \le z.$$

Show by induction on k that

$$\frac{z^k}{k!a_1 \ldots a_k} \le N(a_1, \ldots, a_k; z) \le \frac{(z + a_1 + \cdots + a_k)^k}{k!a_1 \ldots a_k}.$$

[Hint: observe that

$$N(a_1, \ldots, a_k; z) = \sum_{j=0}^{n} N(a_1, \ldots, a_{k-1}; z - ja_k)$$

with $n = [z/a_k]$ and apply the previous exercise.]

15. Deduce from the previous result the following: for $2 \le y \le \sqrt{(\log x)\log\log x}$,

$$\Psi(x, y) = \frac{1}{\pi(y)!} \prod_{p \le y} \left(\frac{\log x}{\log p}\right)\left(1 + O\left(\frac{y^2}{(\log x)(\log y)}\right)\right).$$

This is a result due to Ennola [15].

16. As usual, let $\Phi(x, y)$ be the number of $n \le x$ all of whose prime factors are greater than y. Prove the following analogue of **Buchstab's identity**:

$$\Phi(x, y) = 1 + \sum_{y < p \le x} \sum_{j \ge 1} \Phi(x/p^j, p),$$

for $x, y \ge 1$. Deduce that for $x \ge z \ge y \ge 1$,

$$\Phi(x, y) = \Phi(x, z) + \sum_{y < p \le z} \Phi(x/p, p) + O(x/y).$$

17. Prove that if $\sqrt{x} < y \le x$, then

$$\Phi(x, y) = \pi(x) - \pi(y) + 1.$$

18. By using Buchstab's identity and the prime number theorem in the form

$$\pi(x) = \frac{x}{\log x} + O\left(\frac{x}{\log^2 x}\right),$$

show that, for $x^{1/3} < y \le x^{1/2}$,

$$\Phi(x, y) = \frac{x}{\log x}\{1 + \log(u - 1)\} + O\left(\frac{x}{\log^2 x}\right)$$

with

$$u := \frac{\log x}{\log y}.$$

19. Define the **Buchstab function** $\omega(u)$ recursively, as follows: $u\omega(u) := 1$ for $1 \le u \le 2$ and $u\omega(u) := 1 + \log(u - 1)$ for $2 < u \le 3$. Prove that

$$\Phi(x, y) = \frac{x\omega(u) - y}{\log y} + O\left(\frac{x}{\log^2 y}\right)$$

for $x^{1/3} \le y \le x$.

20. If we define the Buchstab function recursively by

$$u\omega(u) = 1 + \int_1^{u-1} \omega(v)\,dv, \quad u > 2,$$

prove inductively, using Buchstab's identity, that

$$\Phi(x, y) = \frac{x\omega(u) - y}{\log y} + O\left(\frac{x}{\log^2 y}\right)$$

for $x^\varepsilon < y \le x$ with any $\varepsilon > 0$.

21. If $\Psi(x, y)$ denotes the number of $n \le x$ all of whose prime factors are less than y, show that, for any $\varepsilon > 0$,

$$\Psi(x, (\log x)^r) = O(x^{1 - \frac{1}{r} + \varepsilon})$$

for any $r > 0$.

22. Show that, for any $\varepsilon > 0$ and any $r > 0$,

$$\Phi(x, (\log x)^r) = xV((\log x)^r) + O(x^{1 - \frac{1}{r} + \varepsilon}),$$

with $V(\cdot)$ as in Theorem 5.2.1.

23. Let $y = \exp(c \log x / \log \log x)$ for some constant $c > 0$. Let $f_y(n)$ be the number of prime factors of n which are in the interval $[y, x^{1/4}]$. Show that

$$\sum_{\substack{n \le x \\ (n, P_y) = 1}} f_y(n) = xV(y) \left(\log \log \log x + O((\log \log x)/\log x)\right),$$

with $V(y)$ as in Theorem 5.2.1.

24. With notation as in the previous exercise, show that

$$\sum_{\substack{n \le x \\ (n, P_y) = 1}} f_y(n)^2 = xV(y)(\log \log \log x)^2 + O(xV(y) \log \log \log x).$$

25. Using the previous two exercises, show that

$$\sum_{\substack{n \le x, \\ (n, P_y) = 1}} \left(f_y(n) - \log \log \log x\right)^2 = O(xV(y) \log \log \log x).$$

Deduce that Corollary 5.3.4 can be improved to

$$\pi(x) \ll \frac{x \log \log x}{(\log x)(\log \log \log x)}.$$

6

Brun's sieve

Viggo Brun (1885–1978) introduced the sieve that now bears his name in 1915 in the paper [4]. It seems that Jean Merlin [39] had made the first serious attempt to go beyond Eratosthenes. Unfortunately, he was killed in World War I (see [27, p. 33]) and only two of his manuscripts have survived, namely [39, 40]. The latter was prepared for publication by Jacques Hadamard (1865–1963) and published posthumously. Clearly, Brun read Merlin's papers very carefully and was inspired by them. Perhaps he was the only one to have done so. No doubt, this led to his 1915 paper on the subject and later these results were developed into a sophisticated sieve [5].

In his fundamental work, Brun proved that there are infinitely many integers n such that n and $n+2$ have at most nine prime factors. He also showed that all sufficiently large even integers are the sum of two integers, each having at most nine prime factors. These represent tremendous advances towards the twin prime conjecture and the Goldbach conjecture. As a consequence of this work, he deduced that the sum of the reciprocals of the sequence of twin primes converges (see Corollary 5.4.5).

Brun's original papers on the subject were largely ignored. One story reports that Edmund Landau (1877–1938) had not looked at them for eight years, even though they were on his desk. Part of the difficulty lay in the unwieldy notation that Brun had used. Now, 80 years later, the ideas look simple enough and the notation has been streamlined.

As noted before, some of the early results of Brun can be derived using the Eratosthenes sieve and Rankin's trick. But this was not noticed until the paper [51] appeared. However, the later work of Brun on his sieve method cannot be tamed by the elementary methods of the previous chapter. Thus, its study is essential as a viable tool in sieve theory that cannot be ignored.

6.1 Brun's pure sieve

In what follows, we will describe Brun's starting idea for the development of what is now known as **Brun's pure sieve**.

We start with the simple observation that for any positive integers ν and r such that $0 \leq r \leq \nu - 1$,

$$\sum_{k \leq r} (-1)^k \binom{\nu/k} = (-1)^r \binom{\nu - 1}{r}.$$

This is obtained by comparing the coefficients of x^r of both sides of the identity

$$(1-x)^{-1}(1-x)^{\nu} = (1-x)^{\nu-1}.$$

Now let n be a positive integer and let N be the radical of n (that is, the product of the prime divisors of n taken with multiplicity 1). We use the above formula with $\nu = \nu(n)$ to deduce that, for any $0 \leq r \leq \nu(n) - 1$,

$$\sum_{\substack{d|n \\ \nu(d) \leq r}} \mu(d) = \sum_{\substack{d|N \\ \nu(d) \leq r}} \mu(d) = \sum_{k \leq r} (-1)^k \binom{\nu(n)}{r} = (-1)^r \binom{\nu(n) - 1}{r}. \quad (6.1)$$

Let us define the **truncated Möbius function** of d by

$$\mu_r(d) := \begin{cases} \mu(d) & \text{if } \nu(d) \leq r, \\ 0 & \text{if } \nu(d) > r, \end{cases}$$

and let us set

$$\psi_r(n) := \sum_{d|n} \mu_r(d).$$

Then (6.1) can be rewritten as

$$\psi_r(n) = (-1)^r \binom{\nu(n) - 1}{r}, \quad (6.2)$$

a formula that could be viewed as a generalization of the fundamental property of the Möbius function.

From (6.2) we see that $\psi_r(n) \geq 0$ if r is even and $\psi_r(n) \leq 0$ if r is odd. This implies that for any positive integers n and r we have

$$\psi_{2r+1}(n) \leq \sum_{d|n} \mu(d) \leq \psi_{2r}(n). \quad (6.3)$$

We also see that

$$\psi_{2r+1}(n) = \sum_{\substack{d|n \\ \nu(d) \le 2r}} \mu(d) + \sum_{\substack{d|n \\ \nu(d)=2r+1}} \mu(d) = \psi_{2r}(n) + O\left(\sum_{\substack{d|n \\ \nu(d)=2r+1}} |\mu(d)| \right).$$

By combining the two we get that for any positive integers n and r,

$$\sum_{d|n} \mu(d) = \psi_r(n) + O\left(\sum_{\substack{d|n \\ \nu(d)=r+1}} |\mu(d)| \right). \tag{6.4}$$

Brun's ingenious idea was to use $\psi_r(n)$, via (6.4), in the sieve of Eratosthenes so as to reduce the size of the error terms. This is the basic starting point for Brun's pure sieve.

To elucidate the method, we will begin by applying Brun's idea to the problem of obtaining an upper bound for $\Phi(x, z)$, the number of integers $\le x$ that are free of prime factors $< z$. Set, as usual,

$$P_z := \prod_{p<z} p.$$

Then by (6.3) we obtain that, for r even,

$$\Phi(x, z) \le \sum_{n \le x} \sum_{d|(n,P_z)} \mu_r(d)$$

$$= \sum_{d|P_z} \mu_r(d) \left[\frac{x}{d} \right]$$

$$= x \sum_{d|P_z} \frac{\mu_r(d)}{d} + O(z^r), \tag{6.5}$$

since $\mu_r(d) = 0$ unless $\nu(d) \le r$.

We now turn our attention to

$$\sum_{d|P_z} \frac{\mu_r(d)}{d}.$$

By Möbius inversion,

$$\mu_r(d) = \sum_{\delta|d} \mu\left(\frac{d}{\delta}\right) \psi_r(\delta),$$

so that

$$\sum_{d|P_z} \frac{\mu_r(d)}{d} = \sum_{d|P_z} \frac{1}{d} \sum_{\delta|d} \mu\left(\frac{d}{\delta}\right) \psi_r(\delta)$$

$$= \sum_{\delta|P_z} \frac{\psi_r(\delta)}{\delta} \sum_{d|\frac{P_z}{\delta}} \frac{\mu(d)}{d}$$

$$= W(z) \sum_{\delta|P_z} \frac{\psi_r(\delta)}{\phi(\delta)},$$

where

$$W(z) := \prod_{p<z} \left(1 - \frac{1}{p}\right)$$

and ϕ denotes the Euler function. In other words,

$$\sum_{d|P_z} \frac{\mu_r(d)}{d} = W(z) + W(z) \sum_{\substack{\delta|P_z \\ \delta>1}} \frac{\psi_r(\delta)}{\phi(\delta)}. \tag{6.6}$$

We would like to estimate

$$\sum_{\substack{\delta|P_z \\ \delta>1}} \frac{\psi_r(\delta)}{\phi(\delta)}.$$

Let us first observe that, from (6.2),

$$\psi_r(\delta) \le \binom{\nu(\delta)-1}{r},$$

so that the sum under consideration is bounded by

$$\sum_{\substack{\delta|P_z \\ \delta>1}} \binom{\nu(\delta)-1}{r} \frac{1}{\phi(\delta)} \le \sum_{r+1\le m\le\pi(z)} \binom{m-1}{r} \sum_{\substack{\delta|P_z \\ \delta>1 \\ \nu(\delta)=m}} \frac{1}{\phi(\delta)}$$

$$\le \sum_{r+1\le m\le\pi(z)} \binom{m-1}{r} \left(\sum_{p<z} \frac{1}{p-1}\right)^m \frac{1}{m!}$$

$$\le \frac{1}{r!} \sum_{m\ge r+1} \frac{1}{(m-r)!} (\log\log z + c_1)^m$$

$$= \frac{(\log\log z + c_1)^r}{r!} \sum_{m\ge r+1} \frac{1}{(m-r)!} (\log\log z + c_1)^{m-r}$$

$$\le \frac{(\log\log z + c_1)^r}{r!} \exp(\log\log z + c_1).$$

Here we have utilized the elementary estimate that

$$\sum_{p<z} \frac{1}{p} < \log \log z + c_1$$

for some positive constant c_1. Thus

$$\sum_{\substack{\delta|P_z \\ \delta > 1}} \frac{\psi_r(\delta)}{\phi(\delta)} \leq \frac{(\log \log z + c_1)^r}{r!} \exp(\log \log z + c_1). \tag{6.7}$$

We use the well-known estimate

$$\frac{1}{r!} \leq \left(\frac{e}{r}\right)^r$$

to obtain from (6.7) that

$$\sum_{\substack{\delta|P_z \\ \delta > 1}} \frac{\psi_r(\delta)}{\phi(\delta)} \leq c_2 \exp(r - r \log r + r \log \Lambda) \log z, \tag{6.8}$$

where

$$\Lambda := \log \log z + c_1$$

and c_2 is some positive constant.

By combining (6.5), (6.6) and (6.8) we get that

$$\Phi(x, z) \leq x W(z)$$
$$+ x W(z) O \left(\exp \left(r - r \log r + r \log \Lambda \right) \log z \right) + O(z^r). \tag{6.9}$$

Our intention now is to make the $r \log r$ term dominate, to enable us to get a small final error term in (6.9). We choose r to be the nearest even integer to

$$\eta \log \log z$$

for some $\eta = \eta(x, z)$, to be soon specified. With this choice of r, the error terms of (6.9) become

$$x \exp(-\eta (\log \eta)(\log \log z)) + z^{\eta \log \log z},$$

by an elementary calculation. By equating the two terms above we realize that the optimal choice for η is

$$\eta := \frac{\alpha \log x}{(\log z)(\log \log z)} \quad \text{for some } \alpha < 1.$$

In particular, for z satisfying

$$\log z = O\left((\log x)^{1-\varepsilon}\right)$$

for any $0 < \varepsilon < 1$, we find

$$\Phi(x, z) \leq xW(z) + O\left(x \exp\left(-(\log x)^\varepsilon\right)\right).$$

This leads to the estimate

$$\pi(x) = O\left(\frac{x}{(\log x)^{1-\varepsilon}}\right).$$

A more delicate consideration above actually yields that for

$$\log z = \frac{\alpha \log x}{\log \log x}$$

we have

$$\pi(x) = O\left(\frac{x \log \log x}{\log x}\right),$$

a result we obtained by using the sieve of Eratosthenes.

After this brief introduction, we are now in a position to formalize the pure sieve. The starting point is:

Lemma 6.1.1 *Let n, r be positive integers with $r \leq \nu(n)$. Then there exists $|\theta| \leq 1$ such that*

$$\sum_{d|n} \mu(d) = \sum_{\substack{d|n \\ \nu(d) \leq r}} \mu(d) + \theta \sum_{\substack{d|n \\ \nu(d) = r+1}} \mu(d).$$

Proof This is an easy consequence of the combinatorial identities proven at the outset of our discussion. $\square$

Let $\mathcal{A}$ be any set of natural numbers $\leq x$ and let $\mathcal{P}$ be a set of primes. For each prime $p \in \mathcal{P}$, let $\mathcal{A}_p$ be the set of elements of $\mathcal{A}$ which are divisible by p. Let $\mathcal{A}_1 := \mathcal{A}$ and for any squarefree positive integer d composed of primes of $\mathcal{P}$ let $\mathcal{A}_d := \cap_{p|d} \mathcal{A}_p$. Let z be a positive real number and let $P(z) := \prod_{\substack{p \in \mathcal{P} \\ p < z}} p$.

As in the previous chapters, we want to estimate

$$S(\mathcal{A}, \mathcal{P}, z) := \#\left(\mathcal{A} \backslash \cup_{p|P(z)} \mathcal{A}_p\right).$$

We assume that there is a multiplicative function $\omega(\cdot)$ such that, for any d as above,

$$\#\mathcal{A}_d = \frac{\omega(d)}{d} X + R_d \tag{6.10}$$

for some R_d, where

$$X := \#\mathcal{A}.$$

We set

$$W(z) := \prod_{p \mid P(z)} \left(1 - \frac{\omega(p)}{p}\right).$$

Theorem 6.1.2 *(Brun's pure sieve)*
We keep the above setting and we make the additional assumptions that:
1. *$|R_d| \le \omega(d)$ for any squarefree d composed of primes of $\mathcal{P}$;*
2. *there exists a positive constant C such that $\omega(p) < C$ for any $p \in \mathcal{P}$;*
3. *there exist positive constants C_1, C_2 such that*

$$\sum_{\substack{p < z \\ p \in \mathcal{P}}} \frac{\omega(p)}{p} < C_1 \log \log z + C_2.$$

Then

$$S(\mathcal{A}, \mathcal{P}, z) = XW(z)\left(1 + O\left((\log z)^{-A}\right)\right) + O\left(z^{\eta \log \log z}\right)$$

with $A = \eta \log \eta$. In particular, if $\log z \le c \log x / \log \log x$ for a suitable positive constant c sufficiently small, we obtain

$$S(\mathcal{A}, \mathcal{P}, z) = XW(z)(1 + o(1)).$$

Proof   By Lemma 6.1.1 we find that, for any positive integer r,

$$S(\mathcal{A}, \mathcal{P}, z) = \sum_{a \in \mathcal{A}} \sum_{d \mid (a, P(z))} \mu(d)$$

$$= \sum_{a \in \mathcal{A}} \left(\sum_{d \mid (a, P(z))} \mu_r(d) + \theta \sum_{\substack{d \mid (a, P(z)) \\ \nu(d) = r+1}} \mu(d) \right)$$

$$= \sum_{d \mid P(z)} \mu_r(d) \# \mathcal{A}_d + O\left(X \frac{\pi(z)^{r+1}}{(r+1)!}\right).$$

From (6.10), the first hypothesis and the multiplicativity of $\omega(\cdot)$, we get

$$S(\mathcal{A}, \mathcal{P}, z) = X \sum_{d \mid P(z)} \frac{\mu_r(d)\omega(d)}{d} + O\left(\sum_{\substack{d \mid P(z) \\ \nu(d) \le r}} |R_d| \right) + O\left(X \frac{z^{r+1}}{(r+1)!}\right)$$

$$= X \sum_{d \mid P(z)} \frac{\mu_r(d)\omega(d)}{d} + O\left(\left(1 + \sum_{p \mid P(z)} \omega(p)\right)^r \frac{1}{r!}\right)$$

$$+ O\left(X \frac{z^{r+1}}{(r+1)!}\right).$$

Then, by applying the Möbius inversion formula as before, we deduce that

$$S(\mathcal{A}, \mathcal{P}, z) = X \sum_{\delta|P(z)} \frac{\psi_r(\delta)\omega(\delta)}{\delta} \sum_{d|\frac{P(z)}{\delta}} \frac{\mu(d)\omega(d)}{d}$$

$$+ O\left(\left(1 + \sum_{p|P(z)} \omega(p)\right)^r \frac{1}{r!}\right) + O\left(X\frac{z^{r+1}}{(r+1)!}\right).$$

Now let us define

$$\Omega(d) := \prod_{p|d}(p - \omega(p)).$$

The first sum in the expression above becomes

$$XW(z) \sum_{\delta|P(z)} \frac{\psi_r(\delta)\omega(\delta)}{\Omega(\delta)}.$$

By invoking the first and second hypotheses of the theorem, we eventually obtain the desired asymptotic formula. $\square$

6.2 Brun's main theorem

The pure sieve of Brun yields results comparable to the sieve of Eratosthenes. To improve upon this, we will try to use a function

$$f(n) = \sum_{d|n} \mu(d)g(d),$$

with $f(1) = 1$, instead of

$$\sum_{d|n} \mu(d),$$

to initiate the sieving process.

More precisely, let $\mathcal{A}$ be a finite set of integers and let $\mathcal{P}$ be a set of primes. For each prime $p \in \mathcal{P}$ suppose that we are given a subset $\mathcal{A}_p \subseteq \mathcal{A}$. As usual, let $\mathcal{A}_1 := \mathcal{A}$ and for a positive squarefree integer d composed of primes of $\mathcal{P}$ let $\mathcal{A}_d := \cap_{p|d}\mathcal{A}_p$. Let z be a positive real number and let $P(z)$ be the product of the primes in $\mathcal{P}$ that are $< z$. We will use the notation $\mathcal{P}^{(\delta)}$ for the set of primes of $\mathcal{P}$ with the prime divisors of δ removed. Applying Möbius inversion to the sum defining f, we obtain

$$\mu(n)g(n) = \sum_{d|n} \mu\left(\frac{n}{d}\right) f(d),$$

so that

$$\sum_{d|P(z)} \mu(d)g(d)\#\mathcal{A}_d = \sum_{d|P(z)} \#\mathcal{A}_d \sum_{\delta|d} \mu\left(\frac{d}{\delta}\right) f(\delta)$$

$$= \sum_{\delta|P(z)} f(\delta) \sum_{d|\frac{P(z)}{\delta}} \mu(d)\#\mathcal{A}_{d\delta}$$

$$= \sum_{\delta|P(z)} f(\delta) S(\mathcal{A}_\delta, \mathcal{P}^{(\delta)}, z)$$

$$= S(\mathcal{A}, \mathcal{P}, z) + \sum_{\substack{\delta|P(z) \\ \delta>1}} f(\delta) S(\mathcal{A}_\delta, \mathcal{P}^{(\delta)}, z).$$

We can interpret the latter sum as some form of error term. By analyzing this term, Brun discovered an ingenious choice of the function g so as to obtain upper and lower bounds for $S(\mathcal{A}, \mathcal{P}, z)$. We will take up this analysis in what follows.

We begin by rewriting the sum

$$\sum_{\substack{\delta|P(z) \\ \delta>1}} f(\delta) S(\mathcal{A}_\delta, \mathcal{P}^{(\delta)}, z)$$

in a more manageable form. Let $q(\delta)$ denote the smallest prime divisor of δ, where we interpret $q(1)$ as infinity. Let us write each δ as pt with $p = q(\delta)$. Then

$$\sum_{\substack{\delta|P(z) \\ \delta>1}} f(\delta) S(\mathcal{A}_\delta, \mathcal{P}^{(\delta)}, z) = \sum_{p|P(z)} \sum_{\substack{t|\frac{P(z)}{p} \\ p<q(t)}} S(\mathcal{A}_{pt}, \mathcal{P}^{(pt)}, z) f(pt)$$

$$= \sum_{p|P(z)} \sum_{\substack{t|P(z) \\ p<q(t)}} S(\mathcal{A}_{pt}, \mathcal{P}^{(pt)}, z)\left(\sum_{d|t} \mu(d)(g(d)-g(pd))\right).$$

Now write $t = de$ and observe that $p < q(t)$ if and only if $p < q(d)$ and $p < q(e)$. Then

$$\sum_{\substack{\delta|P(z) \\ \delta>1}} f(\delta) S(\mathcal{A}_\delta, \mathcal{P}^{(\delta)}, z) = \sum_{\substack{d|P(z) \\ p<q(d)}} \sum_{p|P(z)} \mu(d)(g(d)-g(pd)) \sum_{\substack{e|\frac{P(z)}{d} \\ p<q(e)}} S(\mathcal{A}_{pde}, \mathcal{P}^{(pde)}, z).$$

If we denote

$$S(\mathcal{A}, \mathcal{P}, z) := \mathcal{A} \setminus \cup_{p|P(z)} \mathcal{A}_p,$$

then a moment's reflection shows that

$$S(\mathcal{A}_{pd}, \mathcal{P}^{(pd)}, p) = \coprod_{\substack{e \mid \frac{P(z)}{d} \\ p < q(e)}} S(\mathcal{A}_{pde}, \mathcal{P}^{(pde)}, z),$$

since any element of the left-hand side must necessarily belong to a unique $\mathcal{A}_{pde}$ for some e with $q(e) > p$ (the right-hand side is a disjoint union). Thus

$$\sum_{\substack{\delta \mid P(z) \\ \delta > 1}} f(\delta) S(\mathcal{A}_\delta, \mathcal{P}^{(\delta)}, z) = \sum_{d \mid P(z)} \sum_{\substack{p \mid P(z) \\ p < q(d)}} \mu(d)(g(d) - g(pd)) S(\mathcal{A}_{pd}, \mathcal{P}^{(pd)}, p).$$

Since $p < q(d)$,

$$S(\mathcal{A}_{pd}, \mathcal{P}^{(pd)}, p) = S(\mathcal{A}_{pd}, \mathcal{P}, p).$$

This proves:

Theorem 6.2.1 *In the above setting and for any function g with $g(1) = 1$, we have*

$$S(\mathcal{A}, \mathcal{P}, z) = \sum_{d \mid P(z)} \mu(d) g(d) \# \mathcal{A}_d - \sum_{d \mid P(z)} \sum_{\substack{p \mid P(z) \\ p < q(d)}} \mu(d)(g(d) - g(pd)) S(\mathcal{A}_{pd}, \mathcal{P}, p).$$

The case when $g(1) := 1$, $g(d) := 0$ for all $d > 1$ gives the famous:

Lemma 6.2.2 *(Buchstab's identity)*
In the above setting,

$$S(\mathcal{A}, \mathcal{P}, z) = \# \mathcal{A} - \sum_{p \mid P(z)} S(\mathcal{A}_p, \mathcal{P}, p).$$

We also deduce:

Theorem 6.2.3 *We keep the setting of Theorem 6.2.1 and let g_U and g_L be two functions such that $g_U(1) = g_L(1) = 1$, satisfying*

$$\mu(d)(g_U(d) - g_U(pd)) \geq 0$$

and

$$\mu(d)(g_L(d) - g_L(pd)) \leq 0$$

for all $d \mid P(z), p < q(d)$. Then

$$\sum_{d \mid P(z)} \mu(d) g_L(d) \# \mathcal{A}_d \leq S(\mathcal{A}, \mathcal{P}, z) \leq \sum_{d \mid P(z)} \mu(d) g_U(d) \# \mathcal{A}_d.$$

Proof The condition on g_U implies that the second sum in Theorem 6.2.1 is non-negative. The upper bound is now immediate. A similar argument applies for g_L. $\square$

We now proceed to choosing functions g_U and g_L satisfying the conditions of Theorem 6.2.3. To simplify the thought, we shall only consider, following Brun, functions g so that $g(d) = 0$ or 1. We decompose the interval $[1, z]$ as

$$2 = z_r < z_{r-1} < \cdots < z_1 < z_0 = z$$

and we define

$$P_{z_n,z} := \prod_{z_n \le p < z} p.$$

We let b be a fixed positive integer and set

$$g_U(d) := \begin{cases} 1 & \text{if } \nu((d, P_{z_n,z})) \le 2b + 2n - 2 \quad \forall n \le r, \\ 0 & \text{otherwise.} \end{cases} \tag{6.11}$$

We claim that the upper bound condition of Theorem 6.2.3 is satisfied. For this we must show that

$$\mu(d) = 1 \quad \text{implies} \quad g_U(d) \ge g_U(pd)$$

and

$$\mu(d) = -1 \quad \text{implies} \quad g_U(d) \le g_U(pd)$$

for any $d \mid P(z)$ and $p < q(d)$. Suppose $\mu(d) = 1$. The only case to consider is $g_U(d) = 0$ and $g_U(pd) = 1$. In this case we have

$$\nu((d, P_{z_m,z})) > 2b + 2m - 2 \text{ for some } m \le r$$

and

$$\nu((pd, P_{z_n,z})) \le 2b + 2n - 2 \quad \forall n \le r.$$

For $n = m$ the second inequality contradicts the first and so the situation cannot arise. Suppose now that $\mu(d) = -1$. The only case to consider is $g_U(pd) = 0$ and $g_U(d) = 1$. In such a case,

$$\nu((d, P_{z_n,z})) \le 2b + 2n - 2 \quad \forall n \le r$$

and

$$\nu((pd, P_{z_m,z})) > 2b + 2m - 2 \text{ for some } m \le r.$$

In the second inequality we have

$$\nu((p, P_{z_m,z})) + \nu((d, P_{z_m,z})) > 2b + 2m - 2,$$

which means

$$z_m \le p < z$$

and

$$2b + 2m - 3 < \nu((d, P_{z_m, z})) \le 2b + 2m - 2.$$

Therefore

$$\nu((d, P_{z_m, z})) = 2b + 2m - 2.$$

Since $p < q(d)$ and $z_m \le p < z$, we deduce

$$\nu(d) = \nu((d, P_{z_m, z})) = 2b + 2m - 2.$$

Hence $\mu(d) = 1$, a contradiction. Notice that, in all cases,

$$(g_U(d) - g_U(pd)) = \mu(d) g_U(d)(1 - g_U(pd)),$$

a fact that will be crucial in our coming discussion.

A similar analysis shows that

$$g_L(d) := \begin{cases} 1 & \text{if } \nu((d, P_{z_n, z})) \le 2b + 2n - 3 \quad \forall n \le r, \\ 0 & \text{otherwise} \end{cases} \tag{6.12}$$

satisfies the condition for g_L in Theorem 6.2.3. We have also that

$$(g_L(d) - g_L(pd)) = -\mu(d) g_L(d)(1 - g_L(pd)).$$

Let p^+ denote the prime of $\mathcal{P}$ that succeeds p. For the choices of g_U and g_L as above, we will prove:

Lemma 6.2.4

$$\sum_{d|P(z)} \mu(d) g_U(d) \frac{\omega(d)}{d} = W(z) \left(1 + \sum_{p < z} \frac{\omega(p) W(p)}{p W(z)} \right.$$

$$\times \sum_{t|P_{p^+, z}} \frac{g_U(t)(1 - g_U(pt))}{t} \omega(t) \Bigg),$$

$$\sum_{d|P(z)} \mu(d) g_L(d) \frac{\omega(d)}{d} = W(z) \left(1 - \sum_{p < z} \frac{\omega(p) W(p)}{p W(z)} \right.$$

$$\times \sum_{t|P_{p^+, z}} \frac{g_L(t)(1 - g_L(pt))}{t} \omega(t) \Bigg).$$

Proof We will establish the identity for g_U, the proof for g_L being similar. Let us observe that for $d|P(z)$,

$$\sum_{p|d}\left(g_U((d,P_{p^+,z}))-g_U((d,P_{p,z}))\right)=1-g_U(d).$$

To see this, note that for $d=1$ the statement is trivial. For $d>1$, let $d=p_1\ldots p_r$ with $p_1<\cdots<p_r$ and $p_i\in\mathcal{P}$. The left-hand side becomes

$$\sum_{i=1}^{r-1}\left(g_U(p_{i+1}\ldots p_r)-g_U(p_i\ldots p_r)\right)+1-g_U(p_r)$$

and the sum telescopes, giving the result. Hence

$$\sum_{d|P(z)}\mu(d)g_U(d)\frac{\omega(d)}{d}=\sum_{d|P(z)}\mu(d)\left(1-\sum_{p|d}(g_U(d,P_{p^+,z})-g_U(d,P_{p,z}))\right)\frac{\omega(d)}{d}$$

$$=W(z)+\sum_{d|P(z)}\sum_{p|d}\mu(d/p)(g_U(d,P_{p^+,z})$$

$$-g_U(d,P_{p,z}))\frac{\omega(d)}{d}.$$

We write $d=\delta pt$, where $\delta|P(p)$ and $t|P_{p^+,z}$, to get

$$\sum_{d|P(z)}\mu(d)g_U(d)\frac{\omega(d)}{d}=W(z)+\sum_{p<z}\frac{\omega(p)}{p}\sum_{\delta|P(p)}\frac{\mu(\delta)\omega(\delta)}{\delta}$$

$$\times\sum_{t|P_{p^+,z}}\mu(t)\frac{g_U(t)-g_U(pt)}{t}\omega(t).$$

We have already noted that $(g_U(t)-g_U(pt))=\mu(t)g_U(t)(1-g_U(pt))$, and so the above becomes

$$\sum_{d|P(z)}\mu(d)g_U(d)\frac{\omega(d)}{d}=W(z)+\sum_{p<z}\frac{\omega(p)}{p}W(p)\sum_{t|P_{p^+,z}}\frac{g_U(t)(1-g_U(pt))}{t}\omega(t),$$

as claimed. This completes the proof. $\square$

We are now ready to prove:

Theorem 6.2.5   *(Brun's sieve)*

We keep the setting of Section 5.1. Suppose that

1. $|R_d| \leq \omega(d)$ *for any squarefree* d *composed of primes of* $\mathcal{P}$*;*
2. *there exists a constant* $A_1 \geq 1$ *such that*

$$0 \leq \frac{\omega(p)}{p} \leq 1 - \frac{1}{A_1};$$

3. *there exist constants* $\kappa > 0$ *and* $A_2 \geq 1$ *such that*

$$\sum_{w \leq p < z} \frac{\omega(p) \log p}{p} \leq \kappa \log \frac{z}{w} + A_2 \quad \text{if} \quad 2 \leq w \leq z.$$

Let b *be a positive integer and let* λ *be a real number satisfying*

$$0 < \lambda e^{1+\lambda} < 1.$$

Then

$$S(\mathcal{A}, \mathcal{P}, z) \leq XW(z) \left\{ 1 + 2 \frac{\lambda^{2b+1} e^{2\lambda}}{1 - \lambda^2 e^{2+2\lambda}} \exp\left((2b+3) \frac{c_1}{\lambda \log z} \right) \right\}$$

$$+ O\left(z^{2b + \frac{2.01}{e^{2\lambda/\kappa} - 1}} \right)$$

and

$$S(\mathcal{A}, \mathcal{P}, z) \geq XW(z) \left\{ 1 - \frac{2\lambda^{2b} e^{2\lambda}}{1 - \lambda^2 e^{2+2\lambda}} \exp\left((2b+2) \frac{c_1}{\lambda \log z} \right) \right\}$$

$$+ O\left(z^{2b - 1 + \frac{2.01}{e^{2\lambda/\kappa} - 1}} \right),$$

where

$$c_1 := \frac{A_2}{2} \left\{ 1 + A_1 \left(\kappa + \frac{A_2}{\log 2} \right) \right\}.$$

Remark 6.2.6   The constants implied by the O-notation do not depend on b or λ. If the first hypothesis of the theorem is replaced by

$$|R_d| \leq L\omega(d),$$

then this changes the theorem by introducing a factor of L into the last error term in each of the upper and lower bound estimates.

Proof With the choices of g_U and g_L as described in (6.11), (6.12), we apply Theorem 6.2.3. We have

$$S(\mathcal{A}, \mathcal{P}, z) \leq \sum_{d|P(z)} \mu(d) g_U(d) \# \mathcal{A}_d$$

$$= \sum_{d|P(z)} \mu(d) g_U(d) \left(\frac{X\omega(d)}{d} + R_d \right).$$

Now we consider

$$\sum_{d|P(z)} \mu(d) g_U(d) \frac{\omega(d)}{d}.$$

By Lemma 6.2.4, this sum is equal to

$$W(z) \left(1 + \sum_{p<z} \frac{\omega(p) W(p)}{p W(z)} \sum_{t|P_{p^+,z}} \frac{g_U(t)(1 - g_U(pt))}{t} \omega(t) \right).$$

The sums in the expression are estimated by

$$\sum_{n=1}^{r} \sum_{z_n \leq p < z_{n-1}} \frac{\omega(p) W(p)}{p W(z)} \sum_{t|P_{p^+,z}} \frac{g_U(t)(1 - g_U(pt))}{t} \omega(t)$$

$$\leq \sum_{n=1}^{r} \frac{W(z_n)}{W(z)} \sum_{z_n \leq p < z_{n-1}} \frac{\omega(p)}{p} \sum_{t|P_{p^+,z}} \frac{g_U(t)(1 - g_U(pt))}{t} \omega(t),$$

since $W(p) \leq W(z_n)$ if $z_n \leq p < z_{n-1}$. Moreover, for each t making a contribution to the innermost sum on the right, we have necessarily that $g_U(t) = 1$ and $g_U(pt) = 0$, which means

$$\nu((t, P_{z_n,z})) \leq 2b + 2n - 2$$

and

$$\nu((pt, P_{z_n,z})) > 2b + 2n - 2,$$

so that, as $t|P_{z_n,z}$,

$$\nu(t) = \nu((t, P_{z_n,z})) = 2b + 2n - 2.$$

Thus the quantity we want to estimate is

$$\leq \sum_{n=1}^{r} \frac{W(z_n)}{W(z)} \sum_{\substack{d \mid P_{z_n,z} \\ \nu(d)=2b+2n-1}} \frac{\omega(d)}{d}$$

$$\leq \sum_{n=1}^{r} \frac{W(z_n)}{W(z)} \cdot \frac{1}{(2b+2n-1)!} \left(\sum_{z_n \leq p < z} \frac{\omega(p)}{p} \right)^{2b+2n-1}.$$

Let λ be a real number satisfying the conditions of the theorem. We will show that $z_1, \ldots, z_r$ can be chosen so that

$$\frac{W(z_n)}{W(z)} \leq e^{2(n\lambda+c)} \quad \text{for} \quad n=1,\ldots,r, \tag{6.13}$$

where

$$c := \frac{c_1}{\log z}.$$

Let us assume this as true for the moment and continue the proof.

We obtain

$$\sum_{z_n \leq p < z} \frac{\omega(p)}{p} \leq \sum_{z_n \leq p < z} \log \left(1 - \frac{\omega(p)}{p}\right)^{-1} = \log \frac{W(z_n)}{W(z)} < 2(n\lambda+c)$$

for $n=1,\ldots,r$. Then it follows that

$$\sum_{n=1}^{r} \frac{W(z_n)}{W(z)} \cdot \frac{1}{(2b+2n-1)!} \left(\sum_{z_n \leq p < z} \frac{\omega(p)}{p} \right)^{2b+2n-1}$$

$$\leq \sum_{n=1}^{r} e^{2n\lambda+2c} \frac{(2n\lambda+2c)^{2b+2n-1}}{(2b+2n-1)!}$$

$$\leq e^{2c}(\lambda+c)^{2b-1} \sum_{n=1}^{r} \frac{(2n/e)^{2n}}{(2n)!} \left(1 + \frac{c}{n\lambda}\right)^{2n} (\lambda e^{1+\lambda})^{2n},$$

since

$$(2b+2n-1)! \geq (2n)!(2n)^{2b-1}.$$

We observe that

$$e^{-n}\frac{n^n}{n!}$$

is decreasing and that

$$\left(1 + \frac{c}{n\lambda}\right)^{2n} \leq e^{2c/\lambda},$$

so that the sum under consideration is at most

$$e^{2c}(\lambda+c)^{2b-1}(2e^{-2})e^{2c/\lambda}\sum_{n=1}^{\infty}(\lambda e^{1+\lambda})^{2n} = \frac{2\lambda^{2b+1}e^{2\lambda}}{1-\lambda^2 e^{2+2\lambda}}\left(1+\frac{c}{\lambda}\right)^{2b-1}e^{2c(1+1/\lambda)}$$

$$\leq \frac{2\lambda^{2b+1}e^{2\lambda}}{1-\lambda^2 e^{2+2\lambda}}e^{(2b-1)c/\lambda+2c+2c/\lambda}$$

$$\leq \frac{2\lambda^{2b+1}e^{2\lambda}}{1-\lambda^2 e^{2+2\lambda}}e^{(2b+3)c/\lambda},$$

since $\lambda < 1$. This proves that

$$\sum_{d|P(z)}\mu(d)g_U(d)\frac{\omega(d)}{d} \leq W(z)\left(1+\frac{2\lambda^{2b+1}e^{2\lambda}}{1-\lambda^2 e^{2+2\lambda}}e^{(2b+3)c/\lambda}\right).$$

An almost identical analysis leads to

$$\sum_{d|P(z)}\mu(d)g_L(d)\frac{\omega(d)}{d} \geq W(z)\left(1-\frac{2\lambda^{2b}e^{2\lambda}}{1-\lambda^2 e^{2+2\lambda}}e^{(2b+2)c/\lambda}\right).$$

We now need to estimate

$$\sum_{d|P(z)}g_U(d)|R_d| \leq \sum_{d|P(z)}g_U(d)\omega(d).$$

The sum is supported on those d satisfying

$$\nu((d, P_{z_n,z})) \leq 2b+2n-2 \qquad \forall n \leq r.$$

We claim that the sum is

$$\leq \left(1+\sum_{p<z}\omega(p)\right)^{2b}\left(1+\sum_{p<z_1}\omega(p)\right)^{2}\cdots\left(1+\sum_{p<z_r}\omega(p)\right)^{2}.$$

To see this, note that if d appears in

$$\sum_{d|P(z)}g_U(d)\omega(d)$$

and $\nu(d) \leq 2b$, then it appears in

$$\left(1+\sum_{p<z}\omega(p)\right)^{2b}.$$

Now suppose $2b < \nu(d) \leq 2b+2$. Not all prime factors of d can be greater than z_1 since $\nu((d, P_{z_1,z})) \leq 2b$. Such a d would therefore appear in

$$\left(1+\sum_{p<z}\omega(p)\right)^{2b}\left(1+\sum_{p<z_1}\omega(p)\right)^{2}.$$

If $2b+2 < \nu(d) \leq 2b+4$, then not all prime factors of d can be greater than z_2 since $\nu((d, P_{z_2,z})) \leq 2b+2$. Such a d would therefore appear in

$$\left(1+\sum_{p<z} \omega(p)\right)^{2b} \left(1+\sum_{p<z_1} \omega(p)\right)^{2} \left(1+\sum_{p<z_2} \omega(p)\right)^{2}.$$

Proceeding in this manner, the claim is established.

By the third hypothesis and partial summation, we deduce

$$\sum_{p<z} \omega(p) \leq A(2\mathrm{li}\, z+3),$$

where $A := \max(\kappa, A_2)$. Inserting this estimate in the above calculation yields

$$\sum_{d|P(z)} g_U(d)\omega(d) \leq (1+A(2\mathrm{li}\, z+3))^{2b} \prod_{n=1}^{r-1}(1+A(2\mathrm{li}\, z_n+3))^{2}.$$

We will now choose our numbers z_n. Let α be a positive real number and define z_n by

$$\log z_n = \mathrm{e}^{-n\alpha} \log z \qquad \text{for} \quad n = 1, \ldots, r.$$

Since $z_r = 2$, we have

$$\log z_{r-1} = \mathrm{e}^{-(r-1)\alpha} \log z > \log 2$$

and

$$\log z_r = \mathrm{e}^{-r\alpha} \log z = \log 2,$$

so that

$$\mathrm{e}^{(r-1)\alpha} < \frac{\log z}{\log 2} = \mathrm{e}^{r\alpha}.$$

Hence, for some absolute constant B,

$$\sum_{d|P(z)} g_U(d)\omega(d) = O\left(\left(\frac{Bz}{\log z}\right)^{2b} \prod_{n=1}^{r-1}\left(\frac{Bz_n \mathrm{e}^{n\alpha}}{\log z}\right)^{2}\right).$$

Let us observe that, since z is sufficiently big,

$$\prod_{n=1}^{r-1} \frac{B\mathrm{e}^{n\alpha}}{\log z} = \frac{B^{r-1}\mathrm{e}^{\frac{1}{2}r(r-1)\alpha}}{(\log z)^{r-1}} = \frac{B^{r-1}(\log z/\log 2)^{\frac{r-1}{2}}}{(\log z)^{r-1}} < 1,$$

$$\frac{B}{\log z} < 1,$$

and

$$\prod_{n=1}^{r-1} z_n^2 = \exp\left(2\sum_{n=1}^{r-1}\log z_n\right) = \exp\left(-2\log z\sum_{n=1}^{r-1} e^{-n\alpha}\right) \le z^{2/(e^\alpha-1)}.$$

Hence the error term is

$$O\left(z^{2b+2/(e^\alpha-1)}\right).$$

It is now time to recall that we must ensure that

$$\frac{W(z_n)}{W(z)} \le e^{2(n\lambda+c)}$$

(see (6.13)). To this end, notice that, by partial summation,

$$\frac{W(w)}{W(z)} \le \exp\left(\kappa\log\frac{\log z}{\log w} + \frac{A_2}{\log w}\left(1+A_1\kappa+\frac{A_1 A_2}{\log 2}\right)\right)$$

for any w, so that

$$\frac{W(z_n)}{W(z)} \le \exp\left(\kappa n\alpha + \frac{2c_1 e^{n\alpha}}{\log z}\right)$$

$$= e^{2c}\exp\left(\kappa\left(n\alpha + \frac{2c_1}{\log z}\cdot\frac{e^{n\alpha}-1}{n}\right)\right)$$

$$\le e^{2c}\exp\left(n\alpha\kappa\left(1+\frac{2c_1 e^\alpha}{\kappa\log 2}\cdot\frac{1}{\log\frac{\log z}{\log 2}}\right)\right).$$

for all $n=1,\dots,r$. Putting

$$\alpha := \frac{2\lambda}{\kappa}\cdot\frac{1}{1+\varepsilon}, \quad \varepsilon := \frac{1}{200 e^{1/\kappa}},$$

we obtain

$$\frac{e^{2\lambda/\kappa}-1}{e^\alpha-1} \le \frac{2.01}{2},$$

so that

$$\sum_{d|P(z)} g_U(d)|R_d| = O\left(z^{2b+\frac{2.01}{e^{2\lambda/\kappa}-1}}\right).$$

The analysis of

$$\sum_{d|P(z)} g_L(d)|R_d|$$

is similar. This completes the proof of the theorem. □

Now let us apply Brun's sieve to the twin prime problem:

Theorem 6.2.7 *As* $x \to \infty$,

$$\#\,\{n \le x : n \text{ and } n+2 \text{ have at most seven prime factors}\} \gg \frac{x}{(\log x)^2}.$$

Proof Consider the sequence

$$\mathcal{A} := \{n(n+2) : \quad n \le x\}.$$

We would like to count the number of elements in this sequence that are free of any prime factors $< z$ for some fixed z. Thus, in the notation of Theorem 6.2.5, $X = x$, $\mathcal{P}$ is the set of rational primes, $\omega(2) = 1$ and $\omega(p) = 2$ for $p > 2$. The hypotheses of the theorem are satisfied with $A_0 = \kappa = 2$, $A_1 = 3$, so that for $b = 1$ we obtain

$$S(\mathcal{A}, \mathcal{P}, z) > \frac{1}{2} x \prod_{2 < p < z} \left(1 - \frac{2}{p}\right) \left\{1 - \frac{2\lambda^2 e^{2\lambda}}{1 - \lambda^2 e^{2+2\lambda}} \exp\left(\frac{4c_1}{\lambda \log z}\right)\right\}$$
$$+ O\left(z^{1 + \frac{2.01}{e^\lambda - 1}}\right).$$

We observe that, for $z \to \infty$,

$$\exp\left(\frac{4c_1}{\lambda \log z}\right) = 1 + O(1/\log z).$$

Now we want to choose λ so that

$$\frac{2\lambda^2 e^{2\lambda}}{1 - \lambda^2 e^{2+2\lambda}} < 1.$$

That is,

$$\lambda e^\lambda < \frac{1}{\sqrt{2 + e^2}}.$$

Using log tables, we easily find that

$$\lambda := \log(1.288)$$

satisfies the above inequalities. For this choice of λ, $7 > 2.01/(e^\lambda - 1)$. Therefore

$$S(\mathcal{A}, \mathcal{P}, z) \gg \frac{x}{\log^2 z} + O\left(z^\theta\right)$$

with $\theta < 8$.

Finally, by choosing $z := x^{1/u}$ with $u < 8$, we deduce that for at least

$$\gg \frac{x}{(\log x)^2}$$

numbers $n \le x$, both n and $n + 2$ have at most seven prime factors. $\square$

6.3 Schnirelman's theorem

In a 1742 letter to Euler, Goldbach conjectured that every integer > 5 can be written as the sum of three primes. This is equivalent to what is now known as **Goldbach's conjecture**, predicting that every positive even integer ≥ 4 can be written as the sum of two primes. Any such integer will be called a **Goldbach number**. In 1770, in his *Meditationes Algebraicae*, Waring formulated another interesting conjecture, now known as **Waring's problem**, saying that every positive integer can be written as the sum of four squares, nine cubes, 19 fourth powers, and so on. In the same year, Lagrange proved his celebrated four square theorem.

In 1909, Hilbert made a significant advance on the Waring problem. For each natural number k, Hilbert showed the existence of a number $g(k)$ so that every natural number can be written as a sum of $g(k)$ k-th powers.

In the 1920s, Hardy and Ramanujan initiated the famous circle method with their attack on the partition function, and, subsequently, they developed the method to handle problems of Goldbach type or Waring's problem. The circle method is an elaborate method that is, in some sense, special to every problem under consideration; a microscopic analysis of the minor arcs is often needed in order to obtain some sort of theorem. This method was subsequently refined, and, by an ingenious use of exponential sums, I. M. Vinogradov proved that every sufficiently large number is the sum of four prime numbers. This theorem is, however, ineffective.

Using Brun's sieve, L. Schnirelman proved in 1939 the following related result:

Theorem 6.3.1   *There exists a positive constant c_0 such that every natural number can be written as a sum of at most c_0 prime numbers.*

The proof of this theorem is exceedingly simple. Apparently, Hardy and Littlewood were astonished at how such a simple argument can yield so much. Moreover, in sharp contrast with the aforementioned Vinogradov's result, Schnirelman's theorem is effective and extremely general in its scope (see [53] where a detailed history of this method is given).

The best constant c_0 that one can obtain by Schnirelman's method has been called **Schnirelman's constant**. Vaughan [71] recently proved that a careful derivation of the argument which we will describe below leads to $c_0 = 27$ and this bound applies to all natural numbers. He also showed [70] that the same method yields an ineffective result that every sufficiently large number is the sum of at most six primes, quite a remarkable achievement for so elementary a method.

Let us describe Schnirelman's method. Let A be a set of non-negative numbers and denote by $A(n)$ the number of positive elements $\leq n$ in the set A. Define the **Schnirelman density** $d(A)$ of the set A by

$$d(A) := \inf_{n \geq 1} \frac{A(n)}{n}.$$

Thus the set E of positive even numbers has $d(E) = 0$, whereas $d(E \cup \{1\}) = 1/2$. If B is another set of non-negative integers, we take

$$A + B := \{a + b : \quad a \in A, \quad b \in B\} \cup A \cup B.$$

We will use the notation $A^{(2)}$ to denote $A + A$, $A^{(3)}$ to denote $A^{(2)} + A$, and so on.

We begin by proving the fundamental property:

Theorem 6.3.2 *Let A, B be sets of non-negative integers. Then*

$$d(A + B) \geq d(A) + d(B) - d(A)d(B).$$

Proof Let n be a positive integer and let $A(n) =: k$. We write

$$0 = a_0 < a_1 < a_2 < \cdots < a_k \leq n$$

for the elements of A.
Clearly,

$$(A + B)(n) \geq A(n) + \sum_{i=0}^{k-1} B(a_{i+1} - a_i - 1) + B(n - a_k).$$

But

$$B(a_{i+1} - a_i - 1) \geq d(B)(a_{i+1} - a_i - 1),$$

thus

$$(A + B)(n) \geq (1 - d(B))A(n) + d(B)n$$

$$\geq (1 - d(B))d(A)n + d(B)n,$$

which proves the theorem. $\square$

Observe that if $d(A) > 1/2$, then $A(n) > n/2$ for all n. Hence the numbers $a, n - a$, as a ranges over all positive $a \in A$ with $a \leq n$, cannot all be distinct, for otherwise we would have $> n$ distinct numbers that are $\leq n$, a contradiction. Therefore, for some $a, a' \in A$, we have $a + a' = n$ and so $A + A = \mathbb{N}$. In other words, if $d(A) > 1/2$, then every natural number can be written as a sum of two elements from A.

Theorem 6.3.2 implies:

Corollary 6.3.3 *Let A be a set of non-negative integers. If $d(A) > 0$, then for some m, $A^{(m)} = \mathbb{N}$.*

Proof If $d(A) = 1$, then the theorem is true with $m = 1$. We may therefore suppose that $d(A) < 1$. Writing the inequality of the theorem as

$$d(A + B) \geq 1 - (1 - d(A))(1 - d(B)),$$

we see easily by induction that

$$d\left(A^{(k)}\right) \geq 1 - (1 - d(A))^k.$$

Now choose k sufficiently large so that

$$(1 - d(A))^k < \frac{1}{2}.$$

Then, for such a k,

$$d\left(A^{(k)}\right) > \frac{1}{2}.$$

This implies that $d\left(A^{(2k)}\right) = 1$ and so the theorem is true with $m = 2k$. $\square$

We would now like to apply these ideas to the study of Goldbach's conjecture and other additive problems. Using Brun's sieve, we will show that the set of Goldbach numbers has positive Schnirelman density. It would then follow from Corollary 6.3.3 that, for some constant c_0, every number can be written as a sum of at most c_0 primes.

We start with a simple application of Brun's sieve:

Theorem 6.3.4 *For any $\alpha \in \mathbb{Z}$, $\alpha \neq 0$, we have*

$$\#\{p \leq x : \ |p + \alpha| \ \text{is} \ \text{prime}\} < \frac{cx}{(\log x)^2} \prod_{p | \alpha} \left(1 - \frac{1}{p}\right)^{-1}$$

for some absolute positive constant c.

Proof We only consider the case $\alpha > 0$, since the case $\alpha < 0$ is almost identical.

We keep the notation of Theorem 6.2.5 and let

$$\mathcal{A} := \{n + \alpha : \quad n \leq x\},$$

$\mathcal{P}$ the set of all primes and $z = z(x)$ a positive real number, to be chosen later. For each prime $p < z$ we would like to ensure $p \nmid n$ and $p \nmid n + \alpha$. Thus, for each prime $p < z$, $p \nmid \alpha$, we have $\omega(p) = 2$, and for $p | \alpha$, we have $\omega(p) = 1$.

The quantity to be estimated in the theorem is clearly

$$\leq z + S(\mathcal{A}, \mathcal{P}, z).$$

Thus we will focus on finding an upper bound for $S(\mathcal{A}, \mathcal{P}, z)$ and then on choosing a suitable z.

In Theorem 6.2.5, we choose λ sufficiently small and $b = 1$ so that

$$S(\mathcal{A}, \mathcal{P}, z) \ll xW(z) + O\left(z^\theta\right)$$

for some constant θ. Moreover, for some constant c_1, we have

$$
\begin{aligned}
W(z) &= \prod_{p < z} \left(1 - \frac{\omega(p)}{p}\right) \\
&= \prod_{\substack{p < z \\ p | \alpha}} \left(1 - \frac{1}{p}\right) \prod_{\substack{p < z \\ p \nmid \alpha}} \left(1 - \frac{2}{p}\right) \\
&= \prod_{\substack{p < z \\ p | \alpha}} \left(1 - \frac{1}{p}\right)\left(1 - \frac{2}{p}\right)^{-1} \prod_{p < z} \left(1 - \frac{2}{p}\right) \\
&\leq \frac{c_1}{(\log z)^2} \prod_{p | \alpha} \left(1 - \frac{1}{p}\right)^{-1}
\end{aligned}
$$

by a simple application of Mertens' theorem. Thus

$$\#\{p \leq x : \quad |p + \alpha| \quad \text{is} \quad \text{prime}\} \ll \frac{x}{(\log z)^2} \prod_{p < z} \left(1 - \frac{2}{p}\right) + z^\theta.$$

Now we choose $z := x^{1/2\theta}$ and deduce the theorem. $\square$

We are ready to prove Schnirelman's Theorem.

Proof of Theorem 6.3.1 Let n be a natural number and let $g(n)$ be the number of ways of writing n as a sum of two primes. Let

$$G(x) := \#\{n \leq x : g(n) \geq 1\}.$$

Then, by the Cauchy–Schwarz inequality,

$$\pi\left(\frac{x}{2}\right)^2 \leq \sum_{\substack{n \leq x \\ g(n) \geq 1}} g(n) \leq G(x)^{1/2} \left(\sum_{\substack{n \leq x \\ g(n) \geq 1}} g^2(n)\right)^{1/2}.$$

This clearly implies that

$$G(x) \gg \frac{x^4}{(\log x)^4 \sum_{\substack{n \leq x \\ g(n) \geq 1}} g^2(n)}.$$

In order to find a lower bound for $G(x)$ it suffices to have an upper bound for

$$\sum_{\substack{n \leq x \\ g(n) \geq 1}} g^2(n).$$

By Theorem 6.3.4 with $b := -n$ and $x := n$, we obtain that, for n such that $g(n) \geq 1$,

$$g(n) < \frac{cn}{(\log n)^2} \prod_{p \mid n}\left(1 - \frac{1}{p}\right)^{-1} \ll \frac{\phi(n)}{(\log n)^2}.$$

Hence

$$\sum_{\substack{n \leq x \\ g(n) \geq 1}} g^2(n) \ll \sum_{1 < n \leq x} \frac{\phi(n)^2}{(\log n)^2} \ll \frac{x^3}{(\log x)^4}.$$

This proves

$$G(x) \gg x.$$

Now let us take $n_1, n_2, \ldots$ to be the sequence of Goldbach numbers. By the above estimate, the sequence

$$1 < n_1 < n_2 < \cdots$$

has positive Schnirelman density. From this we may deduce that there is a positive integer k so that any natural number m can be written as

$$m = \sum_{i \leq k} n_i' + \sum 1$$

where the n_i's are taken from the n_is and in the second sum we have at most k 1s. If the number of 1s is at least 2, we may write the second sum using 2s and 3s. As each n_i' is a sum of two primes, we deduce that m can be written as a sum of $O(k)$ primes. If there is only one 1 in the second sum, we may consider, for $m > 2$, the number $m - 2$ and write

$$m - 2 = \sum_{i \leq k} n'_i + \sum 1.$$

If now there is only one 1, we have

$$m = \sum_{i \leq k} n'_i + 3.$$

Otherwise, we may, as before, group together the 2s and 3s and deduce that m can be written as a sum of at most $O(k)$ primes. This completes the proof. $\square$

This method is capable of considerable refinement and extensive application. First, it is a celebrated **theorem of Mann** that

$$d(A + B) \geq \min(1, d(A) + d(B)).$$

Second, the notion of Schnirelman density can be modified so that the infimum is taken over all $n \geq n_0$ for some sufficiently large n_0. Both Theorem 6.3.2 and Mann's theorem can be established for this modified density. Corollary 6.3.3 would then be altered to read that all sufficiently large numbers can be written as a sum of at most m elements from the set A, where $m = 2k$ and k is the least number so that the modified density $d\left(A^{(k)}\right) > 1/2$. Thirdly, one can prove (using the large sieve inequality) that, for $x \to \infty$,

$$G(x) \sim \frac{1}{2}x,$$

as predicted by Goldbach's conjecture. By Mann's theorem, it would then follow that every sufficiently large number is the sum of at most six primes.

Another application of Schnirelman's theorem that comes to mind is in Waring's problem, namely that for each integer $k \geq 2$ there is a $m(k)$ so that every natural number can be written as a sum of at most $m(k)$ k-th powers. The existence of $m(k)$ was first established by D. Hilbert in two papers written in 1909. The idea of using Schnirelman's theorem to solve Waring's problem was first suggested by U. V. Linnik and his proof was subsequently simplified by L. K. Hua [26]. As we mentioned at the beginning, the theorem for squares was proven by Lagrange in 1770. The corresponding theorem for cubes, namely that $m(3) = 9$, was proven by A. Wieferich in 1909. The result that $m(4) = 19$ was only recently established by R. Balasubramanian and J.-M. Deshouillers. The value of $m(k)$ is now known for all k except $k = 5$. For further details and precise references, we refer the reader to [53].

Elsholtz [14] has discussed the 'inverse' Goldbach problem. Is it possible to find two sets A and B with at least two elements so that $A + B$ is, apart from finitely many exceptions, equal to the set of primes? It is conjectured that no such decomposition exists.

6.4 A theorem of Romanoff

An interesting conjecture of de Polignac asserts that every odd number greater than 3 can be written as the sum of a prime and a power of 2. We relegate to the exercises the construction of counterexamples to this conjecture. However, one can use the above methods to show:

Theorem 6.4.1 *(Romanoff)*
Let $a \geq 2$ be an integer. Then there exists a positive integer r such that every positive integer n can be written as

$$n = p_1 + p_2 + \cdots + p_r + a^{k_1} + a^{k_2} + \cdots + a^{k_r}$$

for some primes $p_1, p_2, \ldots, p_r$ and positive integers $k_1, k_2, \ldots, k_r$.

Remark 6.4.2 *Linnik had established this with $r = 3$.*

Proof We will show that the set of numbers of the form

$$p + a^m,$$

where p is prime, has positive Schnirelman density. By Corollary 6.3.3, this will prove the theorem.

Let us denote by $R(x)$ the number of positive integers $n \leq x$ of the form $p + a^m$, and for each such n set

$$f(n) := \#\{(p, m) : \quad p + a^m = n\}.$$

Then

$$\sum_{n \leq x} f(n) \geq \sum_{p \leq \frac{x}{2}} \sum_{m \leq \frac{\log \frac{x}{2}}{\log a}} 1 \gg x$$

and, by the Cauchy–Schwarz inequality,

$$\sum_{n \leq x} f(n) \leq R(x)^{1/2} \left(\sum_{n \leq x} f^2(n) \right)^{1/2}.$$

It remains to estimate

$$\sum_{n \leq x} f^2(n).$$

First we see that this sum is

$$\leq \# \left\{ (p, m), (p_1, m_1) : p \leq x, p_1 \leq x, m \leq \frac{\log x}{\log a}, m_1 \leq \frac{\log x}{\log a}, p - p_1 = a^{m_1} - a^m \right\}.$$

Now, for a positive integer b, set

$$J(b) := \prod_{p|b} \left(1 + \frac{1}{p} \right)$$

and observe that $J(\cdot)$ is a multiplicative function whose value at b is determined by the prime divisors of b. By Theorem 6.3.4 we infer that, for some positive constant c,

$$\sum_{n \leq x} f^2(n) \leq \frac{cx}{\log^2 x} \sum_{\substack{m, m_1 \leq \frac{\log x}{\log a} \\ m \neq m_1}} J\left(a^{m_1} - a^m\right) + O(x)$$

$$\ll \frac{x}{\log x} \sum_{h \leq \frac{\log x}{\log a}} J\left(a^h - 1\right) + O(x),$$

where the $O(x)$ term represents the contribution from terms with $m = m_1$.

Let $e(d)$ be the exponent of a modulo d. That is, it is the smallest positive number m such that $a^m \equiv 1 \pmod{d}$. If $y := \log x / \log a$, we have

$$\sum_{h \leq y} J\left(a^h - 1\right) = \sum_{h \leq y} \sum_{d | a^h - 1} \frac{\mu^2(d)}{d}$$

$$\leq \sum_{d \leq x} \frac{\mu^2(d)}{d} \sum_{\substack{h \leq y \\ e(d)|h}} 1$$

$$\leq y \sum_{d \leq x} \frac{\mu^2(d)}{d \, e(d)}$$

$$\leq y \sum_{n=1}^{\infty} \frac{s(n)}{n},$$

where

$$s(n) := \sum_{e(d)=n} \frac{\mu^2(d)}{d}.$$

Define

$$S(n) := \sum_{m \le n} s(m) = \sum_{e(d) \le n} \frac{\mu^2(d)}{d}$$

and observe that any d entering into the definition of $S(n)$ must be a squarefree divisor of

$$\prod_{m \le n} (a^m - 1).$$

It is elementary to show (see the exercises) that

$$\sum_{d|n} \frac{\mu^2(d)}{d} \le \frac{6e^\gamma}{\pi^2} \log \log n + O(1).$$

Therefore

$$S(n) \le \frac{6e^\gamma}{\pi^2} \log \log \left(\prod_{m \le n} (a^m - 1) \right) + O(1) \le O(\log n).$$

By partial summation we deduce

$$\sum_{n=1}^{\infty} \frac{s(n)}{n} = \sum_{n=1}^{\infty} \frac{S(n)}{n(n+1)} < \infty.$$

Thus

$$\sum_{h \le y} J\left(a^h - 1\right) \ll y.$$

We conclude that

$$\sum_{n \le x} f^2(n) \ll x,$$

which completes the proof of Romanoff's theorem. $\square$

6.5 Exercises

1. Let k be a natural number and $(a, k) = 1$. Set

$$\mathcal{A} := \{kn + a : n \le x/k\}$$

and

$$\mathcal{P} := \{p \le z : (p, k) = 1\},$$

with usual notation. Let $\pi(x; k, a)$ be the number of primes $p \leq x$ with $p \equiv a(\mathrm{mod}\ k)$. Use Brun's sieve to estimate $S(\mathcal{A}, \mathcal{P}, z)$. Choosing z appropriately, deduce that for some constant $\theta > 0$,

$$\pi(x; k, a) \ll \frac{x}{\phi(k) \log(x/k)}$$

uniformly for $k \leq x^\theta$. (This result is called the **Brun–Titchmarsh theorem**. Sharper results will be derived in later chapters.)

2. Observing that

$$\frac{1}{\phi(k)} = \frac{1}{k} \prod_{p|k} \left(1 - \frac{1}{p}\right)^{-1} \ll \frac{1}{k} \sum_{d|k} \frac{\mu^2(d)}{d},$$

deduce that

$$\sum_{k \leq x} \frac{1}{\phi(k)} \ll \log x.$$

3. Prove that

$$\frac{k}{\phi(k)} = \sum_{d|k} \frac{\mu^2(d)}{\phi(d)}.$$

Deduce that

$$\sum_{k \leq x} \frac{k}{\phi(k)} = Ax + O(\log x),$$

for some positive constant A.

4. Using partial summation, deduce from the previous exercise that there are positive constants A, B such that

$$\sum_{k \leq x} \frac{1}{\phi(k)} = A \log x + B + O\left(\frac{\log x}{x}\right).$$

5. Show that if $d(n)$ is the divisor function, then

$$\sum_{p \leq x} d(p - 1) = O(x).$$

6. Let $\nu(n)$ denote the number of prime factors of n. Show that

$$\sum_{p \leq x} \nu(p - 1) \ll \frac{x \log \log x}{\log x},$$

where the summation is over prime numbers.

7. Let $\Omega(n)$ denote the number of prime factors of n counted with multiplicity. Show that

$$\sum_{p \le x} \Omega(p-1) \ll \frac{x \log \log x}{\log x}.$$

8. Show that

$$\sum_{d|n} \frac{\mu^2(d)}{d} \le \frac{6e^\gamma}{\pi^2} \log \log n + O(1).$$

9. Show that the number of $n \le x$ not having a prime factor $\equiv 3 \pmod 4$ is

$$\ll \frac{x}{\sqrt{\log x}}.$$

10. Let $1 < y \le x$. Show that the number of primes in the interval $[x, x+y]$ is

$$\ll \frac{y}{\log y}.$$

11. Prove that the number of primes in the sequence $n^2 + 1$ with $n \le x$ is

$$\ll \frac{x}{\log x}.$$

12. Show that there are infinitely many odd numbers that cannot be written as $p + 2^m$. [Hint: consider numbers of the form $2^n - 1$.]

13. For a fixed positive integer m, show that the number of solutions of the equation

$$p - 1 = mq$$

with p and q prime $\le x$ is

$$\ll \frac{x}{\phi(m) \log^2(x/m)}.$$

14. Using Brun's sieve, deduce that

$$\pi(x) \ll \frac{x}{\log x}.$$

15. Using the prime number theorem, show that, as x tends to infinity,

$$\sum_{p_n \le x} \frac{p_n - p_{n-1}}{\log p_n} \le (1 + o(1)) \frac{x}{\log x},$$

where the summation is over prime numbers $p_n \le x$ arranged in increasing order. Deduce that

$$\liminf_{n \to \infty} \frac{p_n - p_{n-1}}{\log p_n} \le 1.$$

A long-standing conjecture that the above limit is zero was recently proved by D. Goldston, J. Pintz and C. Yildirim. This represents a significant advance towards the twin prime conjecture (see Section 10.2).

16. Using Brun's sieve, show that the number of solutions of

$$a = p_i - p_j, \quad p_i, p_j \le 2x,$$

where p_i, p_j denote prime numbers, is bounded by

$$\frac{cx}{\log^2 x} \prod_{p|a} \left(1 + \frac{1}{p}\right),$$

for some absolute constant $c > 0$.

17. Let $\delta > 0$ and I denote the interval $[(1-\delta)\log x, (1+\delta)\log x]$. Show that

$$\sum_{a \in I} \prod_{p|a} \left(1 + \frac{1}{p}\right) \ll \delta \log x,$$

where the implied constant is absolute.

18. Let p_n denote the n-th prime. Using the previous two results, show that for some $\delta > 0$,

$$\liminf_{n \to \infty} \frac{p_{n+1} - p_n}{\log p_n} < 1 - \delta.$$

[Hint: if the result is false, then for all n,

$$p_{n+1} - p_n \ge (1 - \delta)\log p_n.$$

Let $p_1 < p_2 < \cdots < p_t$ be the primes in the interval $[x, 2x]$. Clearly,

$$\sum_{n=1}^{t-1}(p_{n+1} - p_n) \le x.$$

Break the sum into two parts according as $p_{n+1} - p_n \in I$ or not.]

19. By observing that the numbers $m!+2, m!+3, \ldots, m!+m$ are all composite for any natural number $m \ge 2$, deduce that for infinitely many n,

$$p_n - p_{n-1} > \frac{\log p_n}{\log\log p_n},$$

where p_n denotes the n-th prime.

20. With p_n denoting the n-th prime, observe that

$$\sum_{p_n \le x}(p_n - p_{n-1}) = (1 + o(1))x.$$

Let

$$G(x) := \max_{p_n \leq x}(p_n - p_{n-1}).$$

Show that

$$G(x) \gg \log x.$$

Deduce that

$$p_n - p_{n-1} \gg \log p_n$$

for infinitely many n.

21. Show that the Schnirelman density of the set of squarefree numbers is strictly greater than $1/2$. Using Mann's theorem, deduce that every natural number can be written as the sum of two squarefree numbers.

22. Let k be a natural number and $(a, k) = 1$. Let $p(k, a)$ denote the least prime $\equiv a(\mathrm{mod}\ k)$. Show that there are positive constants c_1, c_2 such that for k sufficiently large,

$$p(k, a) \geq c_1 \phi(k) \log k$$

for at least $c_2 \phi(k)$ progressions modulo k. [Hint: show that the sum

$$\sum_{(a,k)=1} \pi(x; k, a)(\pi(x; k, a) - 1) \ll \sum_{t \leq x/k} \frac{x}{\log^2 x} \prod_{p|kt}\left(1 + \frac{1}{p}\right),$$

by an application of Brun's sieve. Notice that by the Cauchy–Schwarz inequality,

$$\left(\sum_{(a,k)=1} \pi(x; k, a)\right)^2 \leq \#\{a : \pi(x; k, a) \neq 0\}\left(\sum_{(a,k)=1} \pi(x; k, a)^2\right).$$

Choose $x = c_1 \phi(k) \log k$ for a suitable c_1 to deduce the result.]

7
Selberg's sieve

In the 1940s Atle Selberg discovered a new sieve method in his research on the zeroes of the Riemann zeta function. There he developed his 'mollifier method' that later gave rise to Selberg's sieve, and he used it to show that a positive proportion of the zeroes of the zeta function lie on the critical line $\mathrm{Re}(s) = 1/2$.

The Selberg sieve is essentially combinatorial in structure. As such, it has been generalized to the context of partially ordered sets by R. Wilson [74]. It may be that its versatility has not been fully realized.

7.1 Chebycheff's theorem revisited

We recall that in Section 1.3 of Chapter 1 we used a combinatorial argument of Chebycheff to show that $\pi(x) = O(x/\log x)$. In the subsequent chapters, the use of Turán's sieve, of the Eratosthenes sieve, and even of Brun's pure sieve, led to weaker upper bounds for $\pi(x)$, namely $\pi(x) = O(x/\log\log x)$ (see Exercise 4 of Chapter 4 and Proposition 5.1.1 of Chapter 5) and $\pi(x) = O(x\log\log x/\log x)$ (see Corollary 5.3.4 of Chapter 5). The proof of the latter result was based on the inclusion–exclusion principle expressed in the form

$$\Phi(x, z) = \sum_{d|P_z} \mu(d) \sum_{\substack{n \leq x \\ d|n}} 1, \tag{7.1}$$

and, consequently, on the equation

$$\Phi(x, z) = \sum_{n \leq x} \sum_{d|(n, P_z)} \mu(d), \tag{7.2}$$

113

where P_z is, as usual, the product of all prime numbers $< z$. A refined analysis of (7.2) eventually led to the formula

$$\Phi(x, z) = x \prod_{p<z} \left(1 - \frac{1}{p}\right) + O\left(x(\log z)^2 \exp\left(-\frac{\log x}{\log z}\right)\right) \qquad (7.3)$$

(see Theorem 5.3.3 of Chapter 5) and then to the upper estimate for $\pi(x)$ stated above.

In 1947, Selberg had the brilliant idea of replacing the Möbius function that appears in (7.1) with a quadratic form, optimally chosen so that the resulting estimates are minimal. More precisely, his crucial observation was that for any sequence (λ_d) of real numbers such that

$$\lambda_1 = 1,$$

one has

$$\sum_{d|k} \mu(d) \leq \left(\sum_{d|k} \lambda_d\right)^2 \qquad (7.4)$$

for any k. Using this observation in (7.2) gives

$$\Phi(x, z) \leq \sum_{n \leq x} \left(\sum_{d|(n, P_z)} \lambda_d\right)^2$$

$$= \sum_{n \leq x} \left(\sum_{d_1, d_2|(n, P_z)} \lambda_{d_1} \lambda_{d_2}\right)$$

$$= \sum_{d_1, d_2|P_z} \lambda_{d_1} \lambda_{d_2} \sum_{\substack{n \leq x \\ [d_1, d_2]|n}} 1.$$

By observing that

$$\#\{n \leq x : n \equiv 0 (\mathrm{mod}\ d)\} = \left[\frac{x}{d}\right] = \frac{x}{d} + O(1),$$

we obtain further the inequality

$$\Phi(x, z) \leq x \sum_{d_1, d_2|P_z} \frac{\lambda_{d_1} \lambda_{d_2}}{[d_1, d_2]} + O\left(\sum_{d_1, d_2|P_z} |\lambda_{d_1}||\lambda_{d_2}|\right) \qquad (7.5)$$

in which the first sum is to be viewed as the main term of our estimates and the O-sum as the error term.

Now, for convenience, let us assume that

$$\lambda_d = 0 \quad \text{for any } d > z.$$

This gives

$$\Phi(x, z) \leq x \sum_{d_1, d_2 \leq z} \frac{\lambda_{d_1} \lambda_{d_2}}{[d_1, d_2]} + O\left(\sum_{d_1, d_2 \leq z} |\lambda_{d_1}| |\lambda_{d_2}|\right). \qquad (7.6)$$

We notice that if we also had $|\lambda_d| \leq 1$, then (7.6) would give us an error term of $O\left(z^2\right)$, which, for $z < x$, is smaller than the error term provided by the general sieve of Eratosthenes (see (7.3)). Hence it is reasonable to hope that Selberg's method will give an improvement to our upper estimates of $\Phi(x, z)$, and, consequently, of $\pi(x)$.

Let us estimate the main term in (7.6). The key observation is to view the sum

$$\sum_{d_1, d_2 \leq z} \frac{\lambda_{d_1} \lambda_{d_2}}{[d_1, d_2]}$$

as a quadratic form in $(\lambda_d)_{d \leq z}$, and to seek to minimize this form. Using that

$$[d_1, d_2](d_1, d_2) = d_1 d_2 \qquad (7.7)$$

and that

$$\sum_{\delta \mid d} \phi(\delta) = d \qquad (7.8)$$

(see Exercise 20 of Chapter 1), we can write

$$\sum_{d_1, d_2 \leq z} \frac{\lambda_{d_1} \lambda_{d_2}}{[d_1, d_2]} = \sum_{d_1, d_2 \leq z} \frac{\lambda_{d_1} \lambda_{d_2}}{d_1 d_2}(d_1, d_2)$$

$$= \sum_{d_1, d_2 \leq z} \frac{\lambda_{d_1} \lambda_{d_2}}{d_1 d_2} \sum_{\delta \mid (d_1, d_2)} \phi(\delta)$$

$$= \sum_{\delta \leq z} \phi(\delta) \sum_{\substack{d_1, d_2 \leq z \\ \delta \mid (d_1, d_2)}} \frac{\lambda_{d_1} \lambda_{d_2}}{d_1 d_2}$$

$$= \sum_{\delta \leq z} \phi(\delta) \left(\sum_{\substack{d \leq z \\ \delta \mid d}} \frac{\lambda_d}{d}\right)^2.$$

Therefore, under the transformation

$$u_\delta := \sum_{\substack{d \leq z \\ \delta \mid d}} \frac{\lambda_d}{d}, \qquad (7.9)$$

the initial quadratic form has been diagonalized to

$$\sum_{\delta \leq z} \phi(\delta) u_\delta^2.$$

Our next aim will be to minimize this new diagonal form, if possible.

We recall that the sequence (λ_d) was chosen subject to the constraints that $\lambda_1 = 1$ and $\lambda_d = 0$ for $d > z$. Equation (7.9) tells us that we must also have conditions on (u_δ). This can be seen by using the dual Möbius inversion formula (Theorem 1.2.3). We obtain

$$\frac{\lambda_\delta}{\delta} = \sum_{\delta|d} \mu\left(\frac{d}{\delta}\right) u_d. \tag{7.10}$$

Thus we have the constraints

$$u_\delta = 0 \text{ for any } \delta > z$$

and

$$\sum_{\delta < z} \mu(\delta) u_\delta = 1. \tag{7.11}$$

Now we use (7.11) to write

$$\sum_{\delta \leq z} \phi(\delta) u_\delta^2 = \sum_{\delta \leq z} \phi(\delta) \left(u_\delta - \frac{\mu(\delta)}{\phi(\delta) V(z)}\right)^2 + \frac{1}{V(z)},$$

where

$$V(z) := \sum_{d \leq z} \frac{\mu^2(d)}{\phi(d)}. \tag{7.12}$$

From this expression we see immediately that the form $\sum_{\delta \leq z} \phi(\delta) u_\delta^2$ has a minimal value of $1/V(z)$, attained at

$$u_\delta = \frac{\mu(\delta)}{\phi(\delta) V(z)}.$$

With the above choice of u_δ, and, consequently, with the choice

$$\lambda_\delta = \delta \sum_{\substack{d \leq z \\ \delta|d}} \frac{\mu(d/\delta)\mu(d)}{\phi(d) V(z)} \tag{7.13}$$

of λ_δ, we get

$$\Phi(x, z) \leq \frac{x}{V(z)} + O\left(\sum_{d_1, d_2 \leq z} |\lambda_{d_1}||\lambda_{d_2}|\right).$$

It remains to analyze the O-error term above. Using (7.13) we obtain

$$V(z)\lambda_\delta = \delta \sum_{\substack{d \leq z \\ \delta|d}} \frac{\mu(d/\delta)\mu(d)}{\phi(d)} = \delta \sum_{t \leq \frac{z}{\delta}} \frac{\mu(t)\mu(\delta t)}{\phi(\delta t)}$$

$$= \delta \sum_{\substack{t \leq \frac{z}{\delta} \\ (t,\delta)=1}} \frac{\mu^2(t)\mu(\delta)}{\phi(\delta)\phi(t)} = \mu(\delta) \prod_{p|\delta}\left(1 + \frac{1}{p-1}\right) \sum_{\substack{t \leq \frac{z}{\delta} \\ (t,\delta)=1}} \frac{\mu^2(t)}{\phi(t)}.$$

This implies that

$$|V(z)||\lambda_\delta| \leq |V(z)|,$$

and so

$$|\lambda_\delta| \leq 1 \quad \text{for any } \delta.$$

Putting everything together finally gives:

Theorem 7.1.1

$$\Phi(x, z) \leq \frac{x}{V(z)} + O\left(z^2\right)$$

as $x, z \to \infty$, where $V(z) := \sum_{d \leq z} \dfrac{\mu^2(d)}{\phi(d)}.$

Let us emphasize that, in the end, the sequence (λ_d) was well determined by (7.13) and was not arbitrary. We could have given this definition right from the beginning, but in that way our choice would not have been motivated.

From Theorem 7.1.1 we can deduce now Chebycheff's upper bound for $\pi(x)$:

Corollary 7.1.2

$$\pi(x) \ll \frac{x}{\log x}.$$

Proof We write, as usual,

$$\pi(x) \leq \Phi(x, z) + z$$

for any $z = z(x) \leq x$, and use Theorem 7.1.1 to estimate $\Phi(x, z)$. To do this we need to find a lower bound for $\sum_{d \leq z} \dfrac{\mu^2(d)}{\phi(d)}$ and to choose an appropriate z.
We have

$$\sum_{d \leq z} \frac{\mu^2(d)}{\phi(d)} \geq \sum_{d \leq z} \frac{\mu^2(d)}{d} = \sum_{d \leq z} \frac{1}{d} - {\sum_{d \leq z}}' \frac{1}{d},$$

where the summation $\sum'_{d \leq z}$ is over non-squarefree integers d. We recall that we showed in Chapter 1 that

$$\sum_{d \leq z} \frac{1}{d} = \log z + O(1).$$

By combining this formula with the observation that

$$\sum'_{d \leq z} \frac{1}{d} \leq \frac{1}{4} \sum_{d \leq \frac{z}{4}} \frac{1}{d},$$

we then obtain

$$\sum_{d \leq z} \frac{\mu^2(d)}{\phi(d)} \gg \log z.$$

Hence

$$\pi(x) \ll \frac{x}{\log z} + z^2.$$

Now we choose

$$z := \left(\frac{x}{\log x} \right)^{1/2}$$

and get our desired estimate. □

Remark 7.1.3 The main terms

$$x \prod_{p < z} \left(1 - \frac{1}{p} \right) \quad \text{and} \quad \frac{x}{\displaystyle\sum_{d \leq z} \frac{\mu^2(d)}{\phi(d)}}$$

provided by the sieve of Eratosthenes and Selberg's method, respectively, have the same order of magnitude. They are both $O(x/\log z)$. However, the error term provided by Selberg's method is smaller than the Eratosthenes error term, and this is why we were now able to retrieve Chebycheff's upper estimate for $\pi(x)$.

7.2 Selberg's sieve

In this section we formalize the method illustrated above. Let $\mathcal{A}$ be any finite set of elements and let $\mathcal{P}$ be a set of primes. For each prime $p \in \mathcal{P}$, let $\mathcal{A}_p$ be a subset of $\mathcal{A}$. We denote by d squarefree numbers composed of primes of $\mathcal{P}$. Let $\mathcal{A}_1 := \mathcal{A}$ and for squarefree integers d composed of primes of $\mathcal{P}$, let $\mathcal{A}_d := \cap_{p|d} \mathcal{A}_p$. Let z be a positive real number and set

$$P(z) := \prod_{\substack{p \in \mathcal{P} \\ p < z}} p.$$

Denote by $S(\mathcal{A}, \mathcal{P}, z)$ the number of elements of

$$\mathcal{A} \backslash \cup_{p|P(z)} \mathcal{A}_p.$$

Theorem 7.2.1 *(Selberg's sieve, 1947)*
We keep the above setting and assume that there exist $X > 0$ and a multiplicative function $f(\cdot)$ satisfying $f(p) > 1$ for any prime $p \in \mathcal{P}$, such that for any squarefree integer d composed of primes of $\mathcal{P}$ we have

$$\#\mathcal{A}_d = \frac{X}{f(d)} + R_d \tag{7.14}$$

for some real number R_d. We write

$$f(n) = \sum_{d|n} f_1(d) \tag{7.15}$$

for some multiplicative function $f_1(\cdot)$ that is uniquely determined by f by using the Möbius inversion formula; that is,

$$f_1(n) = \sum_{d|n} \mu(d) f\left(\frac{n}{d}\right).$$

Also, we set

$$V(z) := \sum_{\substack{d \leq z \\ d|P(z)}} \frac{\mu^2(d)}{f_1(d)}.$$

Then

$$S(\mathcal{A}, \mathcal{P}, z) \leq \frac{X}{V(z)} + O\left(\sum_{\substack{d_1, d_2 \leq z \\ d_1, d_2 | P(z)}} |R_{[d_1, d_2]}|\right).$$

To prove this theorem we will need the dual Möbius inversion formula and the following generalization of (7.7):

Lemma 7.2.2 *Let f be a multiplicative function and let d_1, d_2 be positive squarefree integers. Then*

$$f([d_1, d_2]) f((d_1, d_2)) = f(d_1) f(d_2).$$

Proof Exercise. $\square$

Proof of Theorem 7.2.1 Let (λ_d) be any sequence of real numbers such that

$$\lambda_1 = 1$$

and

$$\lambda_d = 0 \ \text{ for any } d > z.$$

For $a \in \mathcal{A}$, let

$$D(a) := \prod_{\substack{p \in \mathcal{P} \\ a \in \mathcal{A}_p}} p,$$

with the convention that $D(a) := 1$ if $a \notin \mathcal{A}_p$ for any $p \in \mathcal{P}$. Then

$$\sum_{\substack{d|P(z) \\ a \in \mathcal{A}_d}} \mu(d) = \sum_{d|(P(z),D(a))} \mu(d) \le \left(\sum_{d|(P(z),D(a))} \lambda_d \right)^2 = \left(\sum_{\substack{d|P(z) \\ a \in \mathcal{A}_d}} \lambda_d \right)^2. \quad (7.16)$$

Now let us look at the cardinality $S(\mathcal{A}, \mathcal{P}, z)$. Using (7.16), we obtain

$$S(\mathcal{A}, \mathcal{P}, z) = \sum_{\substack{a \in \mathcal{A} \\ a \notin \mathcal{A}_p \forall p|P(z)}} 1 = \sum_{d|P(z)} \mu(d) \sum_{a \in \mathcal{A}_d} 1$$

$$= \sum_{a \in \mathcal{A}} \left(\sum_{\substack{d|P(z) \\ a \in \mathcal{A}_d}} \mu(d) \right) \le \sum_{a \in \mathcal{A}} \left(\sum_{\substack{d|P(z) \\ a \in \mathcal{A}_d}} \lambda_d \right)^2$$

$$= \sum_{a \in \mathcal{A}} \left(\sum_{\substack{d_1,d_2|P(z) \\ a \in \mathcal{A}_{[d_1,d_2]}}} \lambda_{d_1} \lambda_{d_2} \right) = \sum_{d_1,d_2 \le z} \lambda_{d_1} \lambda_{d_2} \# \mathcal{A}_{[d_1,d_2]}.$$

By (7.14) the above expression becomes

$$X \sum_{\substack{d_1,d_2 \le z \\ d_1,d_2|P(z)}} \frac{\lambda_{d_1} \lambda_{d_2}}{f([d_1,d_2])} + O\left(\sum_{\substack{d_1,d_2 \le z \\ d_1,d_2|P(z)}} |\lambda_{d_1}||\lambda_{d_2}||R_{[d_1,d_2]}| \right).$$

Here, the first sum is viewed as the main term of our estimates, while the O-sum is viewed as the error term. As in Section 7.1, we also view the main term as a quadratic form in $(\lambda_d)_{d \le z}$, which we seek to bring to a diagonal form and to minimize. By Lemma 7.2.2 and (7.14),

$$
\sum_{\substack{d_1,d_2|P(z)}} \frac{\lambda_{d_1}\lambda_{d_2}}{f([d_1,d_2])} = \sum_{\substack{d_1,d_2 \le z \\ d_1,d_2|P(z)}} \frac{\lambda_{d_1}\lambda_{d_2}}{f(d_1)f(d_2)} f((d_1,d_2))
$$

$$
= \sum_{\substack{d_1,d_2 \le z \\ d_1,d_2|P(z)}} \frac{\lambda_{d_1}\lambda_{d_2}}{f(d_1)f(d_2)} \sum_{\delta|(d_1,d_2)} f_1(\delta)
$$

$$
= \sum_{\substack{\delta \le z \\ \delta|P(z)}} f_1(\delta) \sum_{\substack{d_1,d_2 \le z \\ d_1,d_2|P(z) \\ \delta|(d_1,d_2)}} \frac{\lambda_{d_1}\lambda_{d_2}}{f(d_1)f(d_2)}
$$

$$
= \sum_{\substack{\delta \le z \\ \delta|P(z)}} f_1(\delta) \left(\sum_{\substack{d \le z \\ d|P(z) \\ \delta|d}} \frac{\lambda_d}{f(d)} \right)^2 .
$$

Thus, under the transformation

$$
u_\delta := \sum_{\substack{d \le z \\ d|P(z) \\ \delta|d}} \frac{\lambda_d}{f(d)}, \tag{7.17}
$$

our quadratic form is reduced to the diagonal form

$$
\sum_{\substack{\delta \le z \\ \delta|P(z)}} f_1(\delta) u_\delta^2. \tag{7.18}
$$

The dual Möbius inversion formula enables us to write

$$
\frac{\lambda_\delta}{f(\delta)} = \sum_{\substack{d|P(z) \\ \delta|d}} \mu\left(\frac{d}{\delta}\right) u_d, \tag{7.19}
$$

and so, by recalling that $\lambda_d = 0$ for $d > z$ and $\lambda_1 = 1$, we obtain

$$
u_\delta = 0 \quad \text{for any } \delta > z
$$

and

$$
\sum_{\substack{\delta \le z \\ \delta|P(z)}} \mu(\delta) u_\delta = 1.
$$

Then we can write, as before,

$$\sum_{\substack{\delta \le z \\ \delta \mid P(z)}} f_1(\delta) u_\delta^2 = \sum_{\substack{\delta \le z \\ \delta \mid P(z)}} f_1(\delta) \left(u_\delta - \frac{\mu(\delta)}{f_1(\delta) V(z)} \right)^2 + \frac{1}{V(z)},$$

from which we see immediately that the minimal value of the quadratic form given in (7.18) is $1/V(z)$ and is attained at

$$u_\delta = \frac{\mu(\delta)}{f_1(\delta) V(z)}. \qquad (7.20)$$

We note that here we have used that $f_1(p) = f(p) - 1 > 0$ for any $p \in \mathcal{P}$, hence, by the multiplicativity of $f_1(\cdot)$, that the coefficients $f_1(d)$ appearing in our form are positive.

It remains to analyze the error term

$$O\left(\sum_{\substack{d_1, d_2 \le z \\ d_1, d_2 \mid P(z)}} |\lambda_{d_1}| |\lambda_{d_2}| |R_{[d_1, d_2]}| \right).$$

More precisely, our aim is to find upper bounds for $|\lambda_\delta|$ for all $\delta \le z$, $\delta \mid P(z)$. From (7.17) and (7.19) we obtain that for such δ, we have

$$V(z) \lambda_\delta = f(\delta) \sum_{\substack{d \le z \\ d \mid P(z) \\ \delta \mid d}} \frac{\mu(d/\delta) \mu(d)}{f_1(\delta)}.$$

$$= f(\delta) \sum_{\substack{t \le \frac{z}{\delta} \\ t \mid P(z) \\ (t, \delta) = 1}} \frac{\mu^2(t) \mu(\delta)}{f_1(t) f_1(\delta)}$$

$$= \mu(\delta) \left(\prod_{p \mid \delta} \frac{f(p)}{f_1(p)} \right) \sum_{\substack{t \le \frac{z}{\delta} \\ t \mid P(z) \\ (t, \delta) = 1}} \frac{\mu^2(t)}{f_1(t)}$$

$$= \mu(\delta) \left(\prod_{p \mid \delta} \left(1 + \frac{1}{f_1(p)} \right) \right) \sum_{\substack{t \le \frac{z}{\delta} \\ t \mid P(z) \\ (t, \delta) = 1}} \frac{\mu^2(t)}{f_1(t)}.$$

Therefore

$$|V(z)| |\lambda_\delta| \le |V(z)|,$$

and so

$$|\lambda_\delta| \leq 1.$$

Consequently, the error term becomes

$$O\left(\sum_{\substack{d_1,d_2 \leq z \\ d_1,d_2|P(z)}} |R_{[d_1,d_2]}|\right),$$

and this completes the proof of the theorem. □

In order to use Theorem 7.2.1 we need lower bounds for the quantity $V(z)$. Useful bounds are given as follows.

Lemma 7.2.3 *We keep the notation of Theorem 7.2.1. We let $\tilde{f}(\cdot)$ be the completely multiplicative function defined by $\tilde{f}(p) := f(p)$ for all primes p and we set*

$$\bar{P}(z) := \prod_{\substack{p \notin \mathcal{P} \\ p < z}} p.$$

Then

1. $V(z) \geq \displaystyle\sum_{\substack{\delta \leq z \\ p|\delta \Rightarrow p|P(z)}} \frac{1}{\tilde{f}(\delta)};$

2. $f(\bar{P}(z))V(z) \geq f_1(\bar{P}(z))\displaystyle\sum_{\delta \leq z} \frac{1}{\tilde{f}(\delta)}.$

Proof

1. We want to express the quotients $1/f_1(d)$ appearing in $V(z)$ in terms of the function f. For this, we observe that for d a squarefree integer such that $d|P(z)$ we have

$$\frac{f(d)}{f_1(d)} = \prod_{p|d} \frac{f(p)}{f_1(p)} = \prod_{p|d} \left(1 - \frac{1}{f(p)}\right)^{-1}$$

$$= \prod_{p|d} \sum_{n \geq 0} \frac{1}{f(p)^n} = {\sum_{k}}' \frac{1}{\tilde{f}(k)},$$

where the sum $\displaystyle{\sum_{k}}'$ is over integers k composed of prime divisors of d.
Then

$$V(z) = \sum_{\substack{d \leq z \\ d|P(z)}} \frac{\mu(d)^2}{f(d)} {\sum_{k}}' \frac{1}{\tilde{f}(k)} \geq \sum_{\substack{\delta \leq z \\ p|\delta \Rightarrow p|P(z)}} \frac{1}{\tilde{f}(\delta)}.$$

2. Similarly to part 1, we write

$$\frac{f(\bar{P}(z))}{f_1(\bar{P}(z))} V(z) = \prod_{\substack{p \notin \mathcal{P} \\ p < z}} \left(1 - \frac{1}{f(p)}\right)^{-1} \sum_{\substack{d \le z \\ d | \bar{P}(z)}} \frac{\mu(d)^2}{f(d)} {\sum_k}' \frac{1}{\tilde{f}(k)}$$

$$= \prod_{\substack{p \notin \mathcal{P} \\ p < z}} \left(\sum_{n \ge 0} \frac{1}{f(p)^n}\right) \sum_{\substack{d \le z \\ d | \bar{P}(z)}} \frac{\mu(d)^2}{f(d)} {\sum_k}' \frac{1}{\tilde{f}(k)}$$

$$\ge \sum_{\delta \le z} \frac{1}{\tilde{f}(z)}.$$

□

7.3 The Brun–Titchmarsh theorem and applications

In this section we will use Selberg's sieve to estimate the number of primes $p \le x$ in a given arithmetic progression. More precisely, for any given coprime integers a and k we will estimate

$$\pi(x; k, a) = \#\{p \le x : p \equiv a(\text{mod } k)\}.$$

The problem of studying the asymptotic behaviour of $\pi(x; k, a)$ goes back to Legendre, who conjectured that given $(a, k) = 1$, there are infinitely many primes p such that $p \equiv a(\text{mod } k)$. It is also a particular case of Buniakowski's conjecture concerning the prime values of integral valued polynomials.

In the late 1830s, Dirichlet proved Legendre's conjecture by using important properties of certain generalizations of the Riemann zeta function (known as Dirichlet L-functions). Dirichlet's work on this problem is viewed now as the beginning of analytic number theory. More precisely, Dirichlet showed that the **analytic density** of the set of primes $p \equiv a(\text{mod } k)$ is $1/\phi(k)$, that is, that

$$\lim_{s \to 1} \frac{\displaystyle\sum_{p \equiv a(\text{mod } k)} 1/p^s}{\log 1/(s-1)} = \frac{1}{\phi(k)}.$$

Actually, there is a more natural definition for the density of a set of primes: a set $\mathcal{P}$ of primes is said to have **natural density** δ if the limit

$$\lim_{x \to \infty} \frac{\#\{p \le x : p \in \mathcal{P}\}}{\#\{p \le x\}}.$$

exists and is equal to δ. It can be proven that if $\mathcal{P}$ has natural density δ, then it also has analytic density δ, however the converse is not true in general.

In the case of the set of primes $p \equiv a(\mathrm{mod}\ k)$, the natural density exists and then, by Dirichlet's theorem, we obtain the asymptotic formula

$$\pi(x; k, a) \sim \frac{1}{\phi(k)} \mathrm{li}\, x.$$

Effective asymptotic formulae for $\pi(x; k, a)$ are also known, as follows:

1. For any $N > 0$, there exists $c(N) > 0$ such that, if

$$k \le (\log x)^N,$$

then

$$\pi(x; k, a) = \frac{1}{\phi(k)} \mathrm{li}\, x + O\left(x \exp\left(-c(N)(\log x)^{1/2}\right)\right),$$

uniformly in k (this is known as the **Siegel–Walfisz theorem**);

2. Assuming a generalized Riemann hypothesis (for Dirichlet L-functions) we have that, for any $k \le x$,

$$\pi(x; k, a) = \frac{1}{\phi(k)} \mathrm{li}\, x + O\left(x^{1/2} \log(kx)\right).$$

The implied O-constant does not depend on k.

Thus, unconditionally, the error terms $\left| \pi(x; k, a) - \frac{1}{\phi(k)} \mathrm{li}\, x \right|$ are known (only) in a range of $k < (\log x)^N$. Finally, the Bombieri–Vinogradov theorem mentioned in Section 3.3 of Chapter 3 allows us to control these error terms, on 'average' and unconditionally, in bigger ranges of k. This important theorem will be discussed in detail in Chapter 9.

Our aim in this chapter is more modest than the results mentioned above. We want to obtain an upper bound for $\pi(x; k, a)$. More precisely, we want to prove:

Theorem 7.3.1 *(The Brun–Titchmarsh theorem)*
Let a and k be positive coprime integers and let x be a positive real number such that $k \le x^\theta$ for some $\theta < 1$. Then, for any $\varepsilon > 0$, there exists $x_0 = x_0(\varepsilon) > 0$ such that

$$\pi(x; k, a) \le \frac{(2+\varepsilon)x}{\phi(k) \log(2x/k)}$$

for all $x > x_0$.

Proof We fix a positive real number $z < x$ and observe that

$$\pi(x; k, a) = \pi(z; k, a) + \#\{z \le p \le x : p \equiv a(\text{mod } k)\}$$

$$\le z + \#\{n \le x : n \equiv a(\text{mod } k), n \not\equiv 0(\text{mod } p) \,\forall p < z, (p, k) = 1\}.$$

If we choose

$$\mathcal{A} := \{n \le x : n \equiv a(\text{mod } k)\},$$

$$\mathcal{P} := \{p : (p, k) = 1\},$$

and

$$\mathcal{A}_p := \{n \le x : n \equiv a(\text{mod } k), n \equiv 0(\text{mod } p)\} \text{ for all } p \in \mathcal{P},$$

$$P(z) := \prod_{\substack{p < z \\ (p,k)=1}} p,$$

then we have

$$S(\mathcal{A}, \mathcal{P}, z) = \#\{n \le x : n \equiv a(\text{mod } k), n \not\equiv 0(\text{mod } p) \,\forall p | P(z)\}.$$

Hence once we obtain an upper bound for $S(\mathcal{A}, \mathcal{P}, z)$, we also obtain an upper bound for $\pi(x; k, a)$.

Since the primes $p \in \mathcal{P}$ are coprime to k, we can use the Chinese Remainder Theorem to deduce that

$$\#\mathcal{A}_d = \frac{x}{kd} + O(1)$$

for any squarefree positive integer d composed of primes of $\mathcal{P}$, where

$$\mathcal{A}_d := \cap_{p|d} \mathcal{A}_p.$$

Thus, in the notation of Theorem 7.2.1, we have $X = x/k$, $f(d) = d$, $f_1(d) = \phi(d)$ and $R_d = O(1)$, and so

$$S(\mathcal{A}, \mathcal{P}, z) \le \frac{x}{kV(z)} + O\left(z^2\right),$$

where

$$V(z) := \sum_{\substack{d \le z \\ (d,k)=1}} \frac{\mu^2(d)}{\phi(d)}.$$

Using part 2 of Lemma 7.2.3 together with formula (7.8), we obtain

$$\frac{k}{\phi(k)} V(z) \ge \sum_{\delta \le z} \frac{1}{\delta} \ge \log z + O(1).$$

Hence

$$\pi(x; k, a) \leq z + S(\mathcal{A}, \mathcal{P}, z) \leq \frac{x}{\phi(k)(\log z + O(1))} + O\left(z^2\right).$$

By choosing

$$z := \left(\frac{2x}{k}\right)^{\frac{1}{2} - \varepsilon}$$

for any arbitrary fixed $0 < \varepsilon < 1$, we get the desired upper estimate for $\pi(x; k, a)$. $\square$

Remark 7.3.2 Montgomery and Vaughan [42] have proved the more precise result that

$$\pi(x + y; k, a) - \pi(x; k, a) \leq \frac{2y}{\phi(k) \log(y/k)}$$

for $y > k$. Any further improvement of the constant 2 in the above estimate would imply the non-existence of Siegel zeros, as noted by several authors beginning with Chowla, Motohashi and Siebert. See [47] for a historical discussion.

   Following Erdös, we apply the Brun–Titchmarsh theorem, Brun's sieve and the sieve of Eratosthenes to the problem of finding an asymptotic formula for the number of $n \leq x$ such that $(n, \phi(n)) = 1$. One can prove, using Sylow theorems, that this is a necessary and sufficient condition for any group of order n to be cyclic.

Theorem 7.3.3 *(Erdös)*
The number of $n \leq x$ such that $(n, \phi(n)) = 1$ is

$$\sim \frac{e^{-\gamma} x}{\log \log \log x}$$

as $x \rightarrow \infty$.

   To prove this theorem, we will need the following preliminary results.

Lemma 7.3.4 *Let $0 < \varepsilon < 1$ and let $p < (\log \log x)^{1-\varepsilon}$. Then*

$$\sideset{}{'}\sum_{q} \frac{1}{q} > \frac{c_1 \log \log x}{p} > (\log \log x)^{\varepsilon/2},$$

where the dash indicates that the sum is over primes $q \equiv 1 \pmod{p}$ satisfying $q < x^{1/(\log \log x)^2}$.

Proof Using the Siegel–Walfisz theorem stated above and partial summation, we deduce the assertion of the lemma. $\square$

Lemma 7.3.5 *Let p be any prime. Then*

$$\sum_{\substack{q \le x \\ q \equiv 1 (\text{mod } p)}} \frac{1}{q} < c_2 \left(\frac{\log\log x + \log p}{p} \right).$$

Proof This follows from Theorem 7.3.1 and partial summation. $\square$

Lemma 7.3.6 *Let $0 < \varepsilon < 1$ and let $z \le (\log\log x)^{1+\varepsilon}$. Then the number of $n \le x$ not divisible by any prime $p \le z$ is*

$$(1 + o(1)) \frac{e^{-\gamma} x}{\log z}$$

as $x \to \infty$.

Proof This is a simple application of the sieve of Eratosthenes. $\square$

Lemma 7.3.7 *Let $0 < \varepsilon < 1$ and let $p \le (\log\log x)^{1+\varepsilon}$. Then the number of $n \le x$ having the least prime divisor p is*

$$(1 + o(1)) \frac{e^{-\gamma} x}{p \log p}$$

as $p \to \infty$.

Proof This is a simple consequence of Lemma 7.3.6. $\square$

Proof of Theorem 7.3.3 Let $A(x)$ be the number of $n \le x$ such that $(n, \phi(n)) = 1$. We partition these numbers into sets $\mathcal{A}_p$ according to the smallest prime divisor p of n, and we denote by $A_p(x)$ the number of elements in $\mathcal{A}_p$. Then

$$A(x) = \sum_p A_p(x) = \sum_1 + \sum_2 + \sum_3,$$

where in $\sum_1$ we have $p < (\log\log x)^{1-\varepsilon}$, in $\sum_2$ we have $(\log\log x)^{1-\varepsilon} < p < (\log\log x)^{1+\varepsilon}$, and in $\sum_3$ we have $p > (\log\log x)^{1+\varepsilon}$ for any fixed $0 < \varepsilon < 1$.

Observe that for each prime p, the numbers enumerated by $A_p(x)$ do not have any prime factor $q \equiv 1 \pmod{p}$. By Brun's sieve and Lemma 7.3.4,

$$A_p(x) \ll \frac{x}{p} \prod_{\substack{q \equiv 1 \pmod{p} \\ q < x^{1/(\log\log x)^2}}} \left(1 - \frac{1}{q}\right) \ll \frac{x}{p} \exp\left(-(\log\log x)^{\varepsilon/2}\right).$$

Therefore

$$\sum_1 = o\left(\frac{x}{\log\log\log x}\right).$$

For $\sum_2$ we use Lemma 7.3.7 to deduce that

$$\sum_2 \ll \frac{x}{\log\log\log x} {\sum_p}' \frac{1}{p},$$

where the dash on the sum indicates that $(\log\log x)^{1-\varepsilon} < p < (\log\log x)^{1+\varepsilon}$. We find

$${\sum_p}' \frac{1}{p} \ll \varepsilon,$$

so that

$$\sum_2 \ll \varepsilon \frac{x}{\log\log\log x}.$$

Finally, by the sieve of Eratosthenes,

$$\sum_3 \le (1+o(1)) \frac{e^{-\gamma} x}{(1+\varepsilon)\log\log\log x},$$

since all the prime divisors of an n enumerated by $\sum_3$ are greater than $(\log\log x)^{1+\varepsilon}$. On the other hand,

$$\sum_3 \ge (1+o(1)) \frac{e^{-\gamma} x}{(1+\varepsilon)\log\log\log x} - \sum_{p>y} \frac{x}{p^2} - \sum_{p>y} \sum_{\substack{q \equiv 1 \pmod{p} \\ pq \le x}} \frac{x}{pq},$$

where $y := (\log\log x)^{1+\varepsilon}$. The penultimate sum is easily seen to be

$$o\left(\frac{x}{\log\log x}\right).$$

The final sum is estimated using Lemma 7.3.5. It is

$$\ll x \sum_{p>y} \frac{\log\log x + \log p}{p^2} \ll \frac{x}{(\log\log x)^{\varepsilon}},$$

by partial summation. This completes the proof. $\square$

7.4 Exercises

1. Let f be a multiplicative function and let d_1, d_2 be positive squarefree
    integers. Then

    $$f([d_1, d_2])f((d_1, d_2)) = f(d_1)f(d_2).$$

2. Use the method of Lagrange's multipliers to find the minimal value of
    the form

    $$\sum_{\substack{\delta \leq z \\ \delta | P(z)}} f_1(\delta)u_\delta^2.$$

    defined in (7.18).

3. Show that if a set of primes has natural density δ, then it has analytic
    density and this is also δ.

4. Find an example of a set of primes that has analytic density, but does
    not have natural density.

5. Let a be a positive integer and x a positive real number. Show that, for
    any $A, B > 0$,

    $$\sum_{\substack{\frac{x^A}{(\log x)^B} \leq k \leq x^A (\log x)^B \\ (k,a)=1}} \pi(x; k, a) \ll \frac{x \log \log x}{(\log x)^2}.$$

6. For any natural number n recall that the radical of n is

    $$\operatorname{rad}(n) := \prod_{p|n} p.$$

    Show that the number of $n \leq x$ such that $\operatorname{rad}(n)$ and $\phi(\operatorname{rad}(n))$ are
    coprime is

    $$\sim \frac{e^{-\gamma} x}{\log \log \log x}$$

    as $x \to \infty$. (This gives an asymptotic formula for the number of $n \leq x$
    such that all groups of order n are nilpotent. Since finite abelian groups
    are nilpotent, we infer an asymptotic formula for the number of $n \leq x$
    such that all groups of order n are abelian.) [Hint: modify the method
    of proof of Theorem 7.3.3 to deduce the result.]

7. Using the Siegel–Walfisz theorem, show that

    $$\sum_{\substack{q<z \\ q \equiv 1 (\bmod\ d)}} \frac{1}{q} = \frac{\log \log z}{\phi(d)} + O(1),$$

    where the summation is over primes $q \equiv 1(\bmod\ d)$.

8. In the previous exercise, show that if A is any positive number and $d < (\log z)^A$, then

$$\sum_{\substack{q<z \\ q\equiv 1 (\text{mod } d)}} \frac{1}{q} = \frac{\log\log z}{\phi(d)} + O\left(\frac{\log d}{\phi(d)}\right).$$

9. For any positive integer d, show that

$$\sum_{\substack{q<z \\ q\equiv 1 (\text{mod } d)}} \frac{1}{q} \ll \frac{\log\log z + \log d}{\phi(d)}.$$

10. Let

$$k(n) := \prod_{p|n}(n, p-1).$$

Show that

$$\sum_{n\leq x} \log k(n) = \left(\sum_{d} \frac{\Lambda(d)}{d\phi(d)} + o(1)\right) x \log\log x$$

as $x\to\infty$. ($k(n)$ is an upper bound for the number of non-isomorphic groups of order n all of whose Sylow subgroups are cyclic.) [Hint: write

$$\log k(n) = \sum_{p|n} \sum_{\substack{d|n \\ d|p-1}} \Lambda(d),$$

and analyze the cases $d < \log\log x$ and $d > \log\log x$ separately, using the previous exercises.]

11. Fix $\varepsilon > 0$. Let $H_k(x)$ be the number of $n \leq x$ of the form $n = p_1 p_2 \ldots p_k m$, where all the p_i are less than $(\log\log x)^{1-\varepsilon}$, all the prime divisors of m are greater than $(\log\log x)^{1+\varepsilon}$ and $(m, \phi(m)) = 1$. By modifying the proof of Theorem 7.3.3, show that

$$H_k(x) \sim \frac{e^{-\gamma} x(\log\log\log\log x)^k}{k! \log\log\log x}$$

as $x \to \infty$.

12. By adapting the method of Theorem 7.3.3, show that the number of $n \leq x$ such that $(n, \phi(n))$ is prime is

$$\sim \frac{e^{-\gamma} x \log\log\log\log x}{\log\log\log x}.$$

[Hint: analyze the contribution from squarefree and non-squarefree n separately.]

13. Prove that the number of $n \leq x$ such that $\nu(n, \phi(n)) = k$ is

$$\sim \frac{e^{-\gamma} x (\log \log \log \log x)^k}{k! \log \log \log x}$$

as $x \to \infty$. [Hint: proceed by induction on k and use the previous two exercises.]

14. Prove that

$$\sum_{d \leq z} \frac{\mu^2(d)}{\phi(d)} \geq \log z.$$

[Hint: notice that the sum in question is

$$\sum_{d \leq z} \frac{\mu^2(d)}{d} \prod_{p \mid d} \left(1 - \frac{1}{p}\right)^{-1} = \sum_{\mathrm{rad}(n) \leq z} \frac{1}{n}.]$$

15. For positive integers k, define

$$H_k(x) := \sum_{\substack{d \leq x \\ (d,k)=1}} \frac{\mu^2(d)}{\phi(d)}.$$

Show that

$$H_1(x) = \sum_{t \mid k} \frac{\mu^2(t)}{\phi(t)} H_k(x/t).$$

and deduce that

$$\frac{H_k(x)}{\log x} \geq \prod_{p \mid k} \left(1 - \frac{1}{p}\right).$$

16. Let $\mathcal{A}$ be a set of natural numbers, $\mathcal{P}$ a set of primes, and for squarefree integers d composed of primes of $\mathcal{P}$ let $\mathcal{A}_d$ be the set of elements of $\mathcal{A}$ divisible by d. Suppose that there exists $X > 0$ such that

$$\#\mathcal{A}_d = \frac{X}{d} + R_d$$

for some real number R_d with $|R_d| \leq 1$. Show that in this setting, Theorem 7.2.1 can be sharpened to

$$S(\mathcal{A}, \mathcal{P}, z) \leq \frac{X}{\log z} \prod_{\substack{p < z \\ p \notin \mathcal{P}}} \left(1 - \frac{1}{p}\right)^{-1} + z^2.$$

[Hint: use the previous exercise.]

17. Let k be a natural number. Prove that the number of $n \leq x$ that are coprime to k is

$$\leq \frac{x}{\log z} \prod_{\substack{p \leq z \\ (p,k)=1}} \left(1 - \frac{1}{p}\right)^{-1} + z^2,$$

for any $z \leq x$.

18. Show that for $z \leq x$,

$$\Phi(x, z) \leq \frac{x}{\log z} + z^2,$$

where $\Phi(x, z)$ is the number of $n \leq x$ free of prime factors $< z$.

19. Let k be a natural number and a an integer coprime to k. For $k < y$, show that

$$\pi(x+y; k, a) - \pi(x; k, a) \leq \frac{2y}{\phi(k) \log(y/k)} + O\left(\frac{y}{k \log^2(y/k)}\right),$$

where the implied constant is absolute. (Using the large sieve, Montgomery and Vaughan [42] have shown that the error term above can be eliminated.)

20. Let $\Phi_{odd}(x, z)$ (respectively $\Phi_{even}(x, z)$) denote the number of $n \leq x$ with an odd (respectively even) number of prime factors, counted with multiplicity, free of prime factors $< z$. Suppose that ρ_d is a bounded sequence of real numbers satisfying

$$\sum_{d|n} \mu(d) \leq \sum_{d|n} \rho_d,$$

with $\rho_d = 0$ for $d > z$. Let $\lambda(n)$ denote the **Liouville function**, which is equal to $(-1)^r$, where r is the total number of prime factors of n counted with multiplicity. Assuming that

$$\sum_{n \leq x} \lambda(n) = O(x \exp(-c(\log x)^{1/2}))$$

for some constant $c > 0$ (this assumption is equivalent to a strong form of the prime number theorem), show that, for any $\theta < 1$ and any $z < x^\theta$, we have

$$\Phi_{odd}(x, z) \leq \frac{x}{2} \sum_{d} \frac{\rho_d}{d} + O(x(\log z) \exp(-c_1 (\log x)^{1/2}))).$$

Establish a similar result for $\Phi_{even}(x, z)$. Deduce that

$$\sum_{d \leq \sqrt{x}} \frac{\rho_d}{d} \geq \frac{2 + o(1)}{\log x}.$$

(This example, due to Selberg [59], shows the **parity problem** of the sieve method. As can be seen, the method gives the same bound for $\Phi_{even}(x, \sqrt{x})$ and $\Phi_{odd}(x, \sqrt{x})$, whereas

$$\Phi_{even}(x, \sqrt{x}) = 0, \quad \Phi_{odd}(x, \sqrt{x}) = (1 + o(1))\frac{x}{\log x},$$

as $x \to \infty$.)

8
The large sieve

The sieves discussed in the previous chapters rely on the use of the Möbius function or of its variations (as in the case of Selberg's sieve). They are classified as 'combinatorial sieves'. In this chapter we will discuss a sieve of a completely different nature, called the **large sieve**, which was introduced by Yuri Linnik (1915–72) in 1941 and was subsequently improved by Rényi (1950), Roth (1965), Bombieri (1965), Davenport and Halberstam (1966), Gallagher (1967), and others. We refer the reader to [8, p. 151] for the history. This sieve can be deduced from a beautiful inequality, known as the **large sieve inequality**, which, at a first sight, does not seem to have the big potential that it actually has.

Linnik's original motivation was to attack **Vinogradov's hypothesis** concerning the size of the least quadratic non-residue $n_p \pmod p$. Vinogradov conjectured that

$$n_p = O(p^\varepsilon)$$

for any $\varepsilon > 0$. The generalized Riemann hypothesis implies

$$n_p = O\big((\log p)^2\big).$$

Linnik proved, using his large sieve, that the number of primes $p \le x$ for which $n_p > p^\varepsilon$ is $O(\log \log x)$. Linnik's paper introduced a new theme in analytic number theory that employed ideas from probability theory.

As will be seen in the following chapters, the large sieve has evolved into a powerful tool, its significant application being the Bombieri–Vinogradov theorem. The latter theorem has often served as a substitute for the use of the generalized Riemann hypothesis in certain contexts, and will be discussed in detail in the next chapter.

135

8.1 The large sieve inequality

We begin with a preliminary lemma.

Lemma 8.1.1 *Let $F : [0, 1] \longrightarrow \mathbb{C}$ be a differentiable function with continuous derivative, extended by periodicity to all $\mathbb{R}$ with period 1. Let z be a positive integer. Then*

$$\sum_{\substack{d \leq z}} \sum_{\substack{1 \leq a \leq d \\ (a,d)=1}} \left| F \left(\frac{a}{d} \right) \right| \leq z^2 \int_0^1 |F(\alpha)| \, d\alpha + \int_0^1 |F'(\alpha)| \, d\alpha. \qquad (8.1)$$

Proof Let $d \leq z$, $a \in [1, d] \cap \mathbb{N}$ with $(a, d) = 1$, and $\alpha \in [0, 1]$. Then

$$-F \left(\frac{a}{d} \right) = -F(\alpha) + \int_{\frac{a}{d}}^{\alpha} F'(t) \, dt.$$

By taking absolute value on both sides, this implies

$$\left| F \left(\frac{a}{d} \right) \right| \leq |F(\alpha)| + \int_{\frac{a}{d}}^{\alpha} |F'(t)| \, dt. \qquad (8.2)$$

Now let us fix $\delta > 0$ (to be chosen later) so that the intervals

$$I \left(\frac{a}{d} \right) := \left(\frac{a}{d} - \delta, \frac{a}{d} + \delta \right)$$

are contained in $[0, 1]$. We integrate (8.2) over $I(a/d)$, with respect to α, and obtain

$$2\delta \left| F \left(\frac{a}{d} \right) \right| \leq \int_{I \left(\frac{a}{d} \right)} |F(\alpha)| \, d\alpha + \int_{I \left(\frac{a}{d} \right)} \int_{\frac{a}{d}}^{\alpha} |F'(t)| \, dt \, d\alpha. \qquad (8.3)$$

Since $\alpha \in I(a/d)$ and $t \in [a/d, \alpha]$, we obtain that $t \in I(a/d)$. Hence the right-hand side of the above inequality is

$$\leq \int_{I \left(\frac{a}{d} \right)} |F(\alpha)| \, d\alpha + \int_{I \left(\frac{a}{d} \right)} \int_{I \left(\frac{a}{d} \right)} |F'(t)| \, dt \, d\alpha$$

$$= \int_{I \left(\frac{a}{d} \right)} |F(\alpha)| \, d\alpha + 2\delta \int_{I \left(\frac{a}{d} \right)} |F'(t)| \, dt$$

$$= \int_{I \left(\frac{a}{d} \right)} |F(\alpha)| \, d\alpha + 2\delta \int_{I \left(\frac{a}{d} \right)} |F'(\alpha)| \, d\alpha.$$

In other words,

$$2\delta \left| F \left(\frac{a}{d} \right) \right| \leq \int_{I \left(\frac{a}{d} \right)} |F(\alpha)| \, d\alpha + 2\delta \int_{I \left(\frac{a}{d} \right)} |F'(\alpha)| \, d\alpha. \qquad (8.4)$$

Now we choose

$$\delta := \frac{1}{2z^2}.$$

With this choice, the intervals $I(a/d)$, $1 \leq a \leq d$, $(a,d) = 1$, $d \leq z$ do not overlap (modulo 1). Indeed, let $x \in I(a/d) \cap I(a'/d')$ for $a/d \neq a'/d'$. Then

$$\left| x - \frac{a}{d} \right| < \delta, \quad \left| x - \frac{a'}{d'} \right| < \delta,$$

and so

$$\left| \frac{a}{d} - \frac{a'}{d'} \right| < 2\delta = \frac{1}{z^2}. \tag{8.5}$$

On the other hand, we have

$$\left| \frac{a}{d} - \frac{a'}{d'} \right| = \frac{|ad' - a'd|}{dd'} \neq 0,$$

since if $ad' = a'd$, then, recalling that $(a,d) = (a',d') = 1$, we obtain $d = d'$, which is false. Thus

$$\left| \frac{a}{d} - \frac{a'}{d'} \right| \geq \frac{1}{dd'} \geq \frac{1}{z^2}. \tag{8.6}$$

Putting together (8.5) and (8.6) we are led to a contradiction.

We sum (8.4) over all intervals $I(a/d)$ and get

$$\frac{1}{z^2} \sum_{d \leq z} \sum_{\substack{1 \leq a \leq d \\ (a,d)=1}} \left| F\left(\frac{a}{d} \right) \right| \leq \sum_{I\left(\frac{a}{d}\right)} \int_{I\left(\frac{a}{d}\right)} |F(\alpha)| \, d\alpha + \frac{1}{z^2} \sum_{I\left(\frac{a}{d}\right)} \int_{I\left(\frac{a}{d}\right)} |F'(\alpha)| \, d\alpha$$

$$\leq \int_0^1 |F(\alpha)| \, d\alpha + \frac{1}{z^2} \int_0^1 |F'(\alpha)| \, d\alpha.$$

This completes the proof of the lemma. □

Now let us choose

$$F(\alpha) := \left(\sum_{n \leq x} a_n e(n\alpha) \right)^2,$$

where $(a_n)_{n \geq 1}$ is an arbitrary sequence of complex numbers, x is a positive integer and for a rational number t, $e(t) := \exp(2\pi i t)$.

For simplicity of notation, set

$$S(\alpha) := \sum_{n \leq x} a_n e(n\alpha),$$

hence

$$F(\alpha) = S(\alpha)^2, \quad F'(\alpha) = 2S(\alpha)S'(\alpha).$$

By Lemma 8.1.1 we obtain

$$\sum_{d \leq z} \sum_{\substack{1 \leq a \leq d \\ (a,d)=1}} \left| S\left(\frac{a}{d}\right) \right|^2 \leq z^2 \int_0^1 |S(\alpha)|^2 \, d\alpha + 2 \int_0^1 |S(\alpha)S'(\alpha)| \, d\alpha.$$

We now recall **Parseval's identity** that

$$\int_0^1 \left| \sum_{n \leq x} a_n \mathrm{e}(n\alpha) \right|^2 \, d\alpha = \sum_{n \leq x} |a_n|^2,$$

thus

$$\int_0^1 |S(\alpha)|^2 \, d\alpha = \sum_{n \leq x} |a_n|^2.$$

This implies that

$$\sum_{d \leq z} \sum_{\substack{1 \leq a \leq z \\ (a,d)=1}} \left| S\left(\frac{a}{d}\right) \right|^2 \leq z^2 \sum_{n \leq x} |a_n|^2 + 2 \int_0^1 |S(\alpha)S'(\alpha)| \, d\alpha.$$

For the second term on the right-hand side of the above inequality we use the Cauchy–Schwarz inequality and, once again, Parseval's identity. We obtain

$$\int_0^1 |S(\alpha)S'(\alpha)| \, d\alpha \leq \left(\int_0^1 |S(\alpha)|^2 \, d\alpha \right)^{1/2} \left(\int_0^1 |S'(\alpha)|^2 \, d\alpha \right)^{1/2}$$

$$\leq \left(\sum_{n \leq x} |a_n|^2 \right)^{1/2} \left(\sum_{n \leq x} 4\pi^2 n^2 |a_n|^2 \right)^{1/2}$$

$$\leq 2\pi x \sum_{n \leq x} |a_n|^2.$$

We record this result as:

Theorem 8.1.2 *(The large sieve inequality)*
Let $(a_n)_{n \geq 1}$ be a sequence of complex numbers and let x, z be positive integers. Then

$$\sum_{d \leq z} \sum_{\substack{1 \leq a \leq d \\ (a,d)=1}} \left| \sum_{n \leq x} a_n \mathrm{e}\left(\frac{na}{d}\right) \right|^2 \leq (z^2 + 4\pi x) \sum_{n \leq x} |a_n|^2, \qquad (8.7)$$

where, for a rational number α, $\mathrm{e}(\alpha) := \exp(2\pi i \alpha)$.

Montgomery and Vaughan [42], and Selberg [60] have independently shown that $z^2 + 4\pi x$ can be replaced by $z^2 + x$.

8.2 The large sieve

We want to deduce a sieve method from the inequality described in Theorem 8.1.2. Let $\mathcal{A}$ be a set of positive integers $n \leq x$ and let $\mathcal{P}$ be a set of primes. For each $p \in \mathcal{P}$, suppose that we are given a set $\{w_{1,p}, \ldots, w_{\omega(p),p}\}$ of $\omega(p)$ residue classes modulo p. Let z be a positive real number and denote by $P(z)$ the product of the primes $p \in \mathcal{P}$, $p < z$. We set

$$\mathcal{S}(\mathcal{A}, \mathcal{P}, z) := \{n \in \mathcal{A} : n \not\equiv w_{i,p} \pmod{p} \ \forall 1 \leq i \leq \omega(p), \ \forall p | P(z)\} \quad (8.8)$$

and we denote by $S(\mathcal{A}, \mathcal{P}, z)$ the cardinality of this set.

Theorem 8.2.1 *(The large sieve)*
With the above notation, we have

$$S(\mathcal{A}, \mathcal{P}, z) \leq \frac{z^2 + 4\pi x}{L(z)},$$

where

$$L(z) := \sum_{d \leq z} \mu^2(d) \prod_{p | d} \frac{\omega(p)}{p - \omega(p)}.$$

The (unusual) idea for the proof of Theorem 8.2.1 is to use sums of the form

$$c_d(n) := \sum_{\substack{1 \leq a \leq d \\ (a,d)=1}} e\left(\frac{na}{d}\right), \quad (8.9)$$

where $n, d \in \mathbb{N}$, called **Ramanujan sums**. They have the following interesting properties:

Proposition 8.2.2 *Let d, d' be positive integers. Then*

1. *if $(d, d') = 1$, we have that $c_{dd'}(n) = c_d(n)c_{d'}(n)$;*

2. $c_d(n) = \sum\limits_{D | (d,n)} \mu\left(\dfrac{d}{D}\right) D;$

3. *if $(d, n) = 1$, we have that $c_d(n) = \mu(d)$, that is,*

$$\mu(d) = \sum_{\substack{1 \leq a \leq d \\ (a,d)=1}} e\left(\frac{na}{d}\right).$$

Proof Part 1 of the proposition is left to the reader as an exercise. Part 3 is a straightforward consequence of part 2. We now prove part 2.

Let

$$\tilde{c}_d(n) := \sum_{1 \le a \le d} e\left(\frac{na}{d}\right).$$

On one hand we can write

$$\tilde{c}_d(n) = e\left(\frac{n}{d}\right) \sum_{0 \le a \le d-1} e\left(\frac{na}{d}\right),$$

and we see that this is $e(n/d)\frac{e(n)-1}{e(n/d)-1}$ if $d \nmid n$ and $e(n/d)d$ if $d \mid n$. In other words,

$$\tilde{c}_d(n) = \begin{cases} 0 & \text{if } d \nmid n \\ d & \text{if } d \mid n. \end{cases} \tag{8.10}$$

On the other hand, we can write

$$\tilde{c}_d(n) = \sum_{D \mid d} \sum_{\substack{1 \le a \le d \\ (a,d)=D}} e\left(\frac{na}{d}\right)$$

$$= \sum_{D \mid d} \sum_{\substack{1 \le a_1 \le \frac{d}{D} \\ (a_1, \frac{d}{D})=1}} e\left(\frac{nDa_1}{d}\right) = \sum_{D \mid d} c_{\frac{d}{D}}(n).$$

By using the Möbius inversion formula we deduce that

$$c_d(n) = \sum_{D \mid d} \mu(D) \tilde{c}_{\frac{d}{D}}(n) = \sum_{D \mid d} \mu\left(\frac{d}{D}\right) \tilde{c}_D(n),$$

which, by (8.10), is

$$\sum_{D \mid (d,n)} \mu\left(\frac{d}{D}\right) D.$$

□

Proof of Theorem 8.2.1 First, let us set some notation. Let $d = p_1 \ldots p_t$ be a positive squarefree integer composed of primes dividing $P(z)$. By the Chinese Remainder Theorem, for any $\underline{i} = (i_1, \ldots, i_t)$ with $1 \le i_1 \le \omega(p_1), \ldots, 1 \le i_t \le \omega(p_t)$, there exists a unique integer $w_{\underline{i},d}$ such that

$$w_{\underline{i},d} \equiv w_{i_j, p_j} (\text{mod } p_j) \ \forall 1 \le j \le t.$$

We denote by $\omega(d)$ the number of all possible $w_{\underline{i},d}$ appearing in this fashion (namely, as we vary $\underline{i}$, but keep d fixed). Clearly, $\omega(d)$ is the product of the $\omega(p_i)$'s.

Now let $n \in \mathcal{S}(\mathcal{A}, \mathcal{P}, z)$. This implies that

$$\left(n - w_{\underline{i},d}, d\right) = 1$$

for any d and $\underline{i}$ as above, and so by part 3 of Proposition 8.2.2 we obtain

$$\mu(d) = c_d \left(n - w_{\underline{i},d}\right) = \sum_{\substack{1 \le a \le d \\ (a,d)=1}} e\left(\frac{\left(n - w_{\underline{i},d}\right)a}{d}\right). \qquad (8.11)$$

We sum (8.11) over all indices $\underline{i}$ corresponding to d and over all integers $n \in \mathcal{S}(\mathcal{A}, \mathcal{P}, z)$ and get

$$\mu(d)\omega(d)S(\mathcal{A}, \mathcal{P}, z) = \sum_{\substack{1 \le a \le d \\ (a,d)=1}} \sum_{w_{\underline{i},d}} e\left(\frac{-w_{\underline{i},d}a}{d}\right) \sum_{n \in \mathcal{S}(\mathcal{A},\mathcal{P},z)} e\left(\frac{na}{d}\right). \qquad (8.12)$$

Squaring (8.12) out and applying the Cauchy–Schwarz inequality gives

$$|\mu(d)\omega(d)S(\mathcal{A}, \mathcal{P}, z)|^2 \le \left(\sum_{\substack{1 \le a \le d \\ (a,d)=1}} \left|\sum_{w_{\underline{i},d}} e\left(\frac{-w_{\underline{i},d}a}{d}\right)\right|^2\right)$$

$$\times \left(\sum_{\substack{1 \le a \le d \\ (a,d)=1}} \left|\sum_{n \in \mathcal{S}(\mathcal{A},\mathcal{P},z)} e\left(\frac{na}{d}\right)\right|^2\right).$$

We write the first factor in the above expression as

$$\sum_{\substack{1 \le a \le d \\ (a,d)=1}} \sum_{w_{\underline{i},d}, w_{\underline{i}',d}} e\left(\frac{\left(w_{\underline{i}',d} - w_{\underline{i},d}\right)a}{d}\right) = \sum_{w_{\underline{i},d}, w_{\underline{i}',d}} c_d\left(w_{\underline{i}',d} - w_{\underline{i},d}\right),$$

and, further, by using part 2 of Proposition 8.2.2, as

$$\sum_{w_{\underline{i},d}, w_{\underline{i}',d}} \sum_{D \mid \left(d, w_{\underline{i},d} - w_{\underline{i}',d}\right)} \mu\left(\frac{d}{D}\right) D = \sum_{D \mid d} \sum_{\substack{w_{\underline{i},d}, w_{\underline{i}',d} \\ w_{\underline{i},d} \equiv w_{\underline{i}',d} (\text{mod } D)}} \mu\left(\frac{d}{D}\right) D$$

$$= \sum_{D \mid d} \mu\left(\frac{d}{D}\right) D\omega(d)\omega\left(\frac{d}{D}\right)$$

$$= d\omega(d) \sum_{E \mid d} \frac{\mu(E)\omega(E)}{E}$$

$$= d\omega(d) \prod_{p \mid d} \left(1 - \frac{\omega(p)}{p}\right)$$

$$= \omega(d) \prod_{p \mid d} (p - \omega(p)).$$

This gives us that

$$|\mu(d)\,\omega(d)S(\mathcal{A},\mathcal{P},z)|^2$$

$$\leq \omega(d)\prod_{p|d}(p-\omega(p))\left(\sum_{\substack{1\leq a\leq d\\(a,d)=1}}\left|\sum_{n\in\mathcal{S}(\mathcal{A},\mathcal{P},z)}e\left(\frac{na}{d}\right)\right|^2\right),$$

or, equivalently, that

$$\mu^2(d)S(\mathcal{A},\mathcal{P},z)^2\prod_{p|d}\frac{\omega(p)}{p-\omega(p)}\leq\sum_{\substack{1\leq a\leq d\\(a,d)=1}}\left|\sum_{n\in\mathcal{S}(\mathcal{A},\mathcal{P},z)}e\left(\frac{na}{d}\right)\right|^2. \qquad (8.13)$$

Now we sum (8.13) over $d \leq z$ and we use Theorem 8.1.2 with the sequence $(a_n)_{n\geq1}$ chosen such that a_n is 1 if $n \in \mathcal{S}(\mathcal{A},\mathcal{P},z)$ and 0 otherwise. We obtain

$$\sum_{d\leq z}\mu(d)^2 S(\mathcal{A},\mathcal{P},z)^2\prod_{p|d}\frac{\omega(p)}{p-\omega(p)}\leq\left(z^2+4\pi x\right)S(\mathcal{A},\mathcal{P},z),$$

which, after doing the obvious cancelations, completes the proof of the theorem. $\square$

Remark 8.2.3 When using Theorem 8.2.1, we need lower bounds for the sum $L(z)$. One way of obtaining a lower bound is by considering the summation over primes $p < z$, and not over all integers d; that is,

$$L(z)\geq\sum_{p<z}\frac{\omega(p)}{p-\omega(p)}.$$

In some situations, the lower bound obtained for $L(z)$ in this manner is sufficient, as in the case of applications to the Euclidean algorithm [29].

8.3 Weighted sums of Dirichlet characters

We want to exploit the large sieve inequality in a slightly different direction than the one of the previous section.

We recall that for a positive integer $d \geq 2$, a **(Dirichlet) character** modulo d is a group homomorphism $\chi:(\mathbb{Z}/d\mathbb{Z})^*\longrightarrow\mathbb{C}^*$. One extends χ to all of $\mathbb{Z}$ by setting $\chi(n)=0$ for any integer n not coprime to d. This is a periodic function, whose values are ϕd−th roots of unity. The **trivial** (or **principal**) **character** modulo d, denoted χ_0, is defined by $\chi_0(n)=1$ for all n coprime to d. In the case that χ is non-trivial and has period strictly less than d, we say that it is an **imprimitive** character; otherwise, we say that it is **primitive**, of **conductor** d.

An important result about Dirichlet characters, which will be useful in Chapter 9, is that for any non-trivial character χ modulo d,

$$\sum_{M+1\leq n\leq M+N} \chi(n) \ll d^{\frac{1}{2}} \log d \qquad (8.14)$$

for any M, N. This is known as the **Pólya–Vinogradov inequality**.

There is an important function, called the **Gauss sum**, which brings together the character χ modulo d and the exponential function $\mathrm{e}\left(\frac{\cdot}{d}\right)$:

$$\tau(\chi) := \sum_{1\leq a\leq d} \chi(a)\mathrm{e}\left(\frac{a}{d}\right).$$

One can show that if χ is a primitive character modulo d, then

$$|\tau(\chi)| = d^{1/2} \qquad (8.15)$$

and, for $(n, d) = 1$,

$$\chi(n) = \frac{1}{\tau(\overline{\chi})} \sum_{1\leq a\leq d} \overline{\chi}(a)\mathrm{e}\left(\frac{na}{d}\right), \qquad (8.16)$$

where $\overline{\chi}$ denotes the complex conjugate of χ. We will use these two identities to obtain modified versions of the large sieve inequality.

Theorem 8.3.1 *(First modified large sieve inequality)*
Let $(a_n)_{n\geq 1}$ be a sequence of complex numbers and let x, z be positive integers. Then

$$\sum_{d\leq z} \frac{d}{\phi(d)} \sideset{}{^*}\sum_{\chi} \left| \sum_{n\leq x} a_n\chi(n) \right|^2 \leq (z^2 + 4\pi x) \sum_{n\leq x} |a_n|^2, \qquad (8.17)$$

where the summation $\sideset{}{^*}\sum_{\chi}$ is over primitive characters χ modulo d.

Proof Let $d \leq z$ be fixed and let n be coprime to d. Let χ be a primitive character modulo d. We multiply (8.16) by a_n, sum it over $n \leq x$ and then square it out. By using (8.15) we obtain

$$\left| \sum_{n\leq x} a_n\chi(n) \right|^2 = \frac{1}{d} \left| \sum_{1\leq a\leq d} \overline{\chi}(a) \sum_{n\leq x} a_n\mathrm{e}\left(\frac{an}{d}\right) \right|^2.$$

Now we sum the above identity over all primitive characters χ modulo d and get

$$\sum_{\chi}^{*}\left|\sum_{n\leq x}a_n\chi(n)\right|^2$$

$$=\frac{1}{d}\sum_{\chi}^{*}\left|\sum_{1\leq a\leq d}\overline{\chi}(a)\sum_{n\leq x}a_n\mathrm{e}\left(\frac{an}{d}\right)\right|^2$$

$$\leq\frac{1}{d}\sum_{\chi}\left|\sum_{1\leq a\leq d}\overline{\chi}(a)\sum_{n\leq x}a_n\mathrm{e}\left(\frac{an}{d}\right)\right|^2$$

$$=\frac{1}{d}\sum_{1\leq a\leq d}\sum_{1\leq b\leq d}\left(\sum_{n\leq x}a_n\mathrm{e}\left(\frac{an}{d}\right)\right)\overline{\left(\sum_{n\leq x}a_n\mathrm{e}\left(\frac{bn}{d}\right)\right)}\sum_{\chi}\overline{\chi}(a)\chi(b),$$

where, for a complex number z, we denote by $\overline{z}$ its complex conjugate. By the orthogonality relations (see Exercise 3), $\sum_{\chi}\overline{\chi}(a)\chi(b)$ is $\phi(d)$ if $a\equiv b(\mathrm{mod}\ d)$ and 0 otherwise. Therefore

$$\frac{d}{\phi(d)}\sum_{\chi}^{*}\left|\sum_{n\leq x}a_n\chi(n)\right|^2\leq\sum_{\substack{1\leq a\leq d\\(a,d)=1}}\left|\sum_{n\leq x}a_n\mathrm{e}\left(\frac{an}{d}\right)\right|^2.$$

We sum over $d\leq z$ and apply the large sieve inequality to obtain (8.17). $\square$

An immediate consequence of Theorem 8.3.1 and the Cauchy–Schwarz inequality is:

Corollary 8.3.2 *Let* $(a_n)_{n\geq 1}$, $(b_m)_{m\geq 1}$ *be sequences of complex numbers and let* x,y,z *be positive integers. Then*

$$\sum_{d\leq z}\frac{d}{\phi(d)}\sum_{\chi}^{*}\left|\sum_{n\leq x}\sum_{m\leq y}a_nb_m\chi(nm)\right|$$

$$\leq\left(z^2+4\pi x\right)^{1/2}\left(z^2+4\pi y\right)^{1/2}\left(\sum_{n\leq x}|a_n|^2\right)^{1/2}\left(\sum_{m\leq y}|b_m|^2\right)^{1/2},\quad(8.18)$$

where the summation $\sum_{\chi}^{*}$ *is over primitive characters* χ *modulo* d.

We also want to prove a variation of (8.18) that will be very useful in Chapter 9.

Theorem 8.3.3 *(Second modified large sieve inequality)*
Let $(a_n)_{n\geq 1}, (b_m)_{m\geq 1}$ be sequences of complex numbers and let x, y, z be positive integers. Then

$$
\sum_{d\leq z} \frac{d}{\phi(d)} \sum_{\chi}^{*} \max_{u} \left| \sum_{n\leq x} \sum_{\substack{m\leq y \\ nm\leq u}} a_n b_m \chi(nm) \right|
$$

$$
\ll \left(z^2+x\right)^{1/2} \left(z^2+y\right)^{1/2} \left(\sum_{n\leq x} |a_n|^2\right)^{1/2} \left(\sum_{m\leq y} |b_m|^2\right)^{1/2} \log(2xy), \quad (8.19)
$$

where the summation $\sum_{\chi}^{*}$ is over primitive characters χ modulo d.

Proof Clearly, in order to prove (8.19) we should make use of inequality (8.18). To do this, we have to handle the condition $nm \leq u$. First we observe that, without loss of generality, we can assume $u = k + 1/2$ for some integer $0 \leq k \leq xy$. Now let $d \leq z$ be fixed, let χ be a primitive character modulo d, and let $T > 0$. By Exercise 17 with $\alpha = \log(nm)$ and $\beta = \log u$, we obtain

$$
\sum_{n\leq x} \sum_{\substack{m\leq y \\ nm\leq u}} a_n b_m \chi(nm) = \int_{-T}^{T} A(t, \chi) B(t, \chi) \frac{\sin(t \log u)}{\pi t} \, dt
$$

$$
+ O\left(\frac{1}{T} \sum_{n\leq x} \sum_{m\leq y} \frac{|a_n b_m|}{\left| \log \frac{nm}{u} \right|} \right), \quad (8.20)
$$

where

$$
A(t, \chi) := \sum_{n\leq x} \frac{a_n \chi(n)}{n^{it}}, \quad B(t, \chi) := \sum_{m\leq y} \frac{b_m \chi(m)}{m^{it}}.
$$

We note that, by our assumption on u, we have

$$
\left| \log \frac{nm}{u} \right| \gg \frac{1}{u} \gg \frac{1}{xy}
$$

and

$$
\sin(t \log u) \ll \min\left\{ 1, |t| \log(2xy) \right\}.
$$

Then using the above in (8.20) gives

$$
\left| \sum_{n\leq x} \sum_{\substack{m\leq y \\ nm\leq u}} a_n b_m \chi(nm) \right| \ll \int_{-T}^{T} |A(t, \chi) B(t, \chi)| \min\left\{ \frac{1}{|t|}, \log(2xy) \right\} dt
$$

$$
+ \frac{xy}{T} \sum_{n\leq x} \sum_{m\leq y} |a_n b_m|.
$$

By applying the Cauchy–Schwarz inequality to the last sum, we get that

$$\left| \sum_{\substack{n \leq x \\ nm \leq u}} \sum_{m \leq y} a_n b_m \chi(nm) \right| \ll \int_{-T}^{T} |A(t, \chi) B(t, \chi)| \min \left\{ \frac{1}{|t|}, \log(2xy) \right\} dt$$

$$+ \frac{x^{3/2} y^{3/2}}{T} \left(\sum_{n \leq x} |a_n|^2 \right)^{1/2} \left(\sum_{m \leq y} |b_m|^2 \right)^{1/2}.$$

We take the maximum over u and then sum over χ and $d \leq z$ to obtain

$$\sum_{d \leq z} \frac{d}{\phi(d)} \sum_{\chi}^{*} \max_{u} \left| \sum_{n \leq x} \sum_{\substack{m \leq y \\ nm \leq u}} a_n b_m \chi(nm) \right|$$

$$\ll \sum_{d \leq z} \frac{d}{\phi(d)} \sum_{\chi}^{*} \int_{-T}^{T} |A(t, \chi) B(t, \chi)| \min \left\{ \frac{1}{|t|}, \log(2xy) \right\} dt$$

$$+ \frac{x^{3/2} y^{3/2}}{T} \sum_{d \leq z} \frac{d}{\phi(d)} \sum_{\chi}^{*} \left(\sum_{n \leq x} |a_n|^2 \right)^{1/2} \left(\sum_{m \leq y} |b_m|^2 \right)^{1/2}. \tag{8.21}$$

Now we use the Cauchy–Schwarz inequality, Corollary 8.3.2 and Exercise 15 to obtain upper bounds for the first term on the right-hand side of (8.21):

$$\sum_{d \leq z} \frac{d}{\phi(d)} \sum_{\chi}^{*} \int_{-T}^{T} |A(t, \chi) B(t, \chi)| \min \left\{ \frac{1}{|t|}, \log(2xy) \right\} dt$$

$$\leq \left(\sum_{d \leq z} \frac{d}{\phi(d)} \sum_{\chi}^{*} \left| \sum_{n \leq x} a_n \chi(n) \right|^2 \right)^{1/2} \left(\sum_{d \leq z} \frac{d}{\phi(d)} \sum_{\chi}^{*} \left| \sum_{m \leq y} b_m \chi(m) \right|^2 \right)^{1/2}$$

$$\times \int_{-T}^{T} \min \left\{ \frac{1}{|t|}, \log(2xy) \right\} dt$$

$$\leq (z^2 + 4\pi x)^{1/2} (z^2 + 4\pi y)^{1/2} \left(\sum_{n \leq x} |a_n|^2 \right)^{1/2} \left(\sum_{m \leq y} |b_m|^2 \right)^{1/2}$$

$$\times (\log T + \log(2xy)). \tag{8.22}$$

For the second term on the right-hand side of (8.21) we observe that there are $\phi(d)$ characters modulo d (see Exercise 4), hence

$$\frac{x^{3/2} y^{3/2}}{T} \sum_{d \leq z} \frac{d}{\phi(d)} \sum_{\chi}^{*} \left(\sum_{n \leq x} |a_n|^2 \right)^{1/2} \left(\sum_{m \leq y} |b_m|^2 \right)^{1/2}$$

$$\leq \frac{x^{3/2} y^{3/2} z^2}{T} \left(\sum_{n \leq x} |a_n|^2 \right)^{1/2} \left(\sum_{m \leq y} |b_m|^2 \right)^{1/2}. \tag{8.23}$$

Putting together (8.21), (8.22) and (8.23), and choosing $T := x^{3/2}y^{3/2}$ gives the desired inequality. $\square$

8.4 An average result

In this section we deduce an average result of Barban, Davenport and Halberstam about the behaviour of the error term in the prime number theorem for primes in an arithmetic progression.

Let x be a positive real number and a, d integers with $(a, d) = 1$. By partial summation, in order to study $\pi(x; d, a)$ it is actually enough to study the function

$$\psi(x; d, a) := \sum_{\substack{n \leq x \\ (a,n)=1}} \Lambda(n), \qquad (8.24)$$

where $\Lambda(\cdot)$ is the von Mangoldt function. Hence our focus in this section will be on the function $\psi(x; d, a)$ instead of $\pi(x; d, a)$.

Theorem 8.4.1 *(The Barban–Davenport–Halberstam theorem)*
For any $A > 0$ and for any z satisfying $x/(\log x)^A \leq z \leq x$, we have

$$\sum_{d \leq z} \sum_{\substack{1 \leq a \leq d \\ (a,d)=1}} \left| \psi(x; d, a) - \frac{x}{\phi(d)} \right|^2 \leq xz \log x.$$

Proof We start by rewriting the error term $\psi(x; d, a) - x/\phi(d)$ in a more convenient form. Namely, for any $d \leq z$ and $1 \leq a \leq d$ with $(a, d) = 1$, we observe that

$$\psi(x; d, a) = \frac{1}{\phi(d)} \sum_{\chi \pmod{d}} \overline{\chi}(a)\psi(x, \chi),$$

where the summation $\sum_{\chi \pmod{d}}$ is over (not necessarily primitive) Dirichlet characters χ modulo d, and where

$$\psi(x, \chi) := \sum_{n \leq x} \Lambda(n)\chi(n). \qquad (8.25)$$

Let χ_0 denote the trivial character modulo d. We observe that

$$\psi(x; d, a) - \frac{x}{\phi(d)} = \frac{1}{\phi(d)} \sum_{\substack{\chi \pmod{d} \\ \chi \neq \chi_0}} \overline{\chi}(a)\psi(x, \chi) + \frac{\psi(x, \chi_0) - x}{\phi(d)}. \qquad (8.26)$$

By using the orthogonality relations, we see that

$$\sum_{\substack{1 \le a \le d \\ (a,d)=1}} \left| \sum_{\substack{\chi (\text{mod } d) \\ \chi \neq \chi_0}} \overline{\chi}(a) \psi(x, \chi) \right|^2 = \phi(d) \sum_{\substack{\chi (\text{mod } d) \\ \chi \neq \chi_0}} |\psi(x, \chi)|^2 . \tag{8.27}$$

Now let us remark that if the character $\chi \neq \chi_0$ modulo d is induced by some primitive character χ_1 modulo d_1, then

$$\psi(x, \chi) = \psi(x, \chi_1) + O\left((\log x)(\log d)\right). \tag{8.28}$$

Indeed, we have

$$\psi(x, \chi_1) - \psi(x, \chi) = \sum_{\substack{p^k \le x \\ p \mid d}} \chi_1\left(p^k\right) \log p \ll \sum_{\substack{p \le x \\ p \mid d}} \left[\frac{\log x}{\log p} \right] \log p \ll (\log x)(\log d).$$

Hence, by using (8.26), (8.27) and (8.28), we get

$$\sum_{d \le z} \sum_{\substack{1 \le a \le d \\ (a,d)=1}} \left| \psi(x; d, a) - \frac{x}{\phi(d)} \right|^2$$

$$\ll \sum_{d \le z} \frac{1}{\phi(d)} \sum_{\substack{\chi (\text{mod } d) \\ \chi \neq \chi_0}} |\psi(x, \chi)|^2 + \sum_{d \le z} \frac{1}{\phi(d)} |\psi(x, \chi_0) - x|^2$$

$$\ll \sum_{d \le z} \frac{1}{\phi(d)} \sum_{\substack{\chi (\text{mod } d) \\ \chi \neq \chi_0}} |\psi(x, \chi_1)|^2 + \sum_{d \le z} \frac{1}{\phi(d)} |\psi(x) - x|^2 + z(\log z)^2 (\log x)^2,$$

where $\psi(x)$ is defined in (1.3) of Chapter 1.

Let us remark that the second term above can be estimated using the following form of the prime number theorem:

$$\psi(x) = x + O\left(x \exp\left(-c\sqrt{\log x}\right)\right) \tag{8.29}$$

for some $c > 0$ (see [8, Chapter 18]). Thus

$$\sum_{d \le z} \frac{1}{\phi(d)} |\psi(x) - x|^2 \ll \frac{x^2 \log z}{(\log x)^A} \ll xz \log x,$$

where $A > 0$ is as in the statement of the theorem.

Since the term $z(\log z)^2 (\log x)^2$ is negligible compared to the desired final estimate of $xz \log x$, we now see that in order to prove the theorem it suffices to show that

$$\sum_{d \leq z} \frac{1}{\phi(d)} \sum_{\substack{\chi(\text{mod } d) \\ \chi \neq \chi_0}} |\psi(x, \chi_1)|^2 \ll xz \log x. \tag{8.30}$$

For each character $\chi \neq \chi_0$ modulo d and its associated primitive character χ_1 modulo d_1, let us write $d = d_1 k$ for some positive integer k. We have

$$\sum_{d \leq z} \frac{1}{\phi(d)} \sum_{\substack{\chi(\text{mod } d) \\ \chi \neq \chi_0}} |\psi(x, \chi_1)|^2$$

$$= \sum_{d_1 \leq z} \sum_{\chi_1(\text{mod } d_1)} |\psi(x, \chi_1)|^2 \sum_{k \leq \frac{z}{d_1}} \frac{1}{\phi(d_1 k)}$$

$$= \sum_{d \leq z} \sideset{}{^*}\sum_{\chi} |\psi(x, \chi)|^2 \sum_{k \leq \frac{z}{d}} \frac{1}{\phi(dk)},$$

where the summation $\sum_{\chi}^{*}$ is over primitive characters χ modulo d.

We observe that the innermost sum satisfies (see Exercise 15)

$$\sum_{k \leq \frac{z}{d}} \frac{1}{\phi(dk)} \ll \frac{1}{\phi(d)} \log \frac{2z}{d}.$$

Therefore, in order to prove (8.30) it is enough to show that

$$\sum_{d \leq z} \frac{1}{\phi(d)} \log\left(\frac{2z}{d}\right) \sideset{}{^*}\sum_{\chi} |\psi(x, \chi)|^2 \ll xz \log x. \tag{8.31}$$

To do this, we remark that, by the first modified large sieve inequality applied to the sequence $a_n := \Lambda(n)$, we have

$$\sum_{d \leq z} \frac{d}{\phi(d)} \sideset{}{^*}\sum_{\chi} |\psi(x, \chi)|^2 \ll (z^2 + x) x \log x, \tag{8.32}$$

since $\sum_{n \leq x} \Lambda(n)^2 \ll x \log x$ (see Exercise 13 of Chapter 1).

We will now use this estimate in a slightly indirect manner. More precisely, let $D = D(x)$ be a parameter to be chosen later and such that $1 < D \leq z$, and let us divide the interval $(D, z]$ into dyadic subintervals $(U, 2U]$ with $U := z/2^k$, where k are integers running from 1 to $\log(z/D)/\log 2$. By (8.32) together with partial summation, we are led to

$$\sum_{U<d\leq 2U} \frac{1}{\phi(d)} \log\left(\frac{2z}{d}\right) \sum_{\chi}^* |\psi(x,\chi)|^2$$

$$\ll \left(\frac{x^2}{U} + Ux\right)(\log x)\log\left(\frac{2z}{U}\right)$$

for each interval $(U, 2U]$. Now we sum over all $U = z/2^k$ and obtain

$$\sum_{D<d\leq z} \frac{1}{\phi(d)} \log\left(\frac{2z}{d}\right) \sum_{\chi}^* |\psi(x,\chi)|^2 \ll \frac{x^2}{D}(\log x)^2 + zx\log x. \qquad (8.33)$$

If we choose

$$D := (\log x)^{A+1} \qquad (8.34)$$

and if we recall that $x/(\log x)^A \leq z \leq x$, then (8.33) becomes

$$\sum_{D<d\leq z} \frac{1}{\phi(d)} \log\left(\frac{2z}{d}\right) \sum_{\chi}^* |\psi(x,\chi)|^2 \ll zx\log x. \qquad (8.35)$$

In order to complete the proof of (8.30), it remains to show that

$$\sum_{2\leq d\leq D} \frac{1}{\phi(d)} \log\left(\frac{2z}{d}\right) \sum_{\chi}^* |\psi(x,\chi)|^2 \ll zx\log x. \qquad (8.36)$$

With D as in (8.34) and for any non-trivial character χ modulo d such that $d \leq D$, the Siegel–Walfisz theorem stated in Section 7.3 of Chapter 7 gives that

$$\psi(x,\chi) \ll x\exp\left(-c\sqrt{\log x}\right) \qquad (8.37)$$

for some positive constant $c = c(A)$ (see Exercise 12). Therefore

$$\sum_{2\leq d\leq D} \frac{1}{\phi(d)} \log\left(\frac{2z}{d}\right) \sum_{\chi}^* |\psi(x,\chi)|^2$$

$$\ll D(\log z)x^2 \exp\left(-2c\sqrt{\log x}\right)$$

$$\ll \frac{x^2}{(\log x)^A} \ll zx\log x.$$

This proves (8.36) and also completes the proof of the theorem. $\square$

8.5 Exercises

1. (Parseval's identity) Let $(a_n)_{n \geq 1}$ be complex numbers and x be a positive integer. Show that

$$\int_0^1 \left| \sum_{n \leq x} a_n e(n\alpha) \right|^2 d\alpha = \sum_{n \leq x} |a_n|^2,$$

   where $e(n\alpha) = \exp(2\pi i n\alpha)$.

2. Let d, d', n be positive integers. Show that, if $(d, d') = 1$, then

$$c_{dd'}(n) = c_d(n)c_{d'}(n),$$

   where $c_d(n)$ is defined in (8.9).

3. (The orthogonality relations) Let χ denote Dirichlet characters modulo d. Then

$$\sum_\chi \overline{\chi(a)}\chi(b) = \begin{cases} \phi(d) & \text{if} \quad a \equiv b \pmod{d} \\ 0 & \text{if} \quad a \not\equiv b \pmod{d} \end{cases}$$

   and

$$\sum_{n \pmod{d}} \chi(n) = \begin{cases} \phi(d) & \text{if} \quad \chi = \chi_0 \\ 0 & \text{if} \quad \chi \neq \chi_0. \end{cases}$$

4. Show that there are exactly $\phi(d)$ Dirichlet characters modulo d.

5. Let χ be a Dirichlet character modulo d. If χ is imprimitive, show that there exist a proper factor d_1 of d and a primitive character χ_1 modulo d_1 such that

$$\chi(n) = \begin{cases} \chi_1(n) & \text{if} \quad (n, d) = 1 \\ 0 & \text{if} \quad (n, d) > 1. \end{cases}$$

6. Let χ be a primitive character modulo d. Show that, for any integer n,

$$\chi(n)\tau(\overline{\chi}) = \sum_{1 \leq a \leq d} \overline{\chi}(a) e\left(\frac{na}{d}\right).$$

   Deduce that

$$|\chi(n)|^2 |\tau(\overline{\chi})|^2 = \sum_{1 \leq a_1 \leq d} \sum_{1 \leq a_2 \leq d} \overline{\chi}(a_1)\chi(a_2) e\left(\frac{n(a_1 - a_2)}{d}\right).$$

7. Use the previous exercise to show that, if χ is a primitive character modulo d, then

$$|\tau(\chi)| = d^{1/2}.$$

8. (The Pólya–Vinogradov inequality) Let χ be a non-trivial character modulo d and let M, N be positive integers. Show that

$$\sum_{M+1\leq n\leq M+N} \chi(n) \ll d^{1/2}\log d.$$

9. Let x be a positive real number and let a, d be positive integers such that $d \leq x, (a, d) = 1$. Show that

$$\psi(x; d, a) = \frac{1}{\phi(d)} \sum_{\chi(\mathrm{mod}\ d)} \overline{\chi}(a)\psi(x, \chi),$$

where the summation $\sum_{\chi(\mathrm{mod}\ d)}$ is over Dirichlet characters modulo d and where $\psi(x, \chi)$ is defined in (8.25).

10. With x, a, d as above, show that

$$\psi(x; d, a) - \frac{x}{\phi(d)} = \frac{1}{\phi(d)} \sum_{\substack{\chi(\mathrm{mod}\ d)\\ \chi\neq\chi_0}} \overline{\chi}(a)\psi(x, \chi) + \frac{\psi(x, \chi_0) - x}{\phi(d)}.$$

11. Let x be a positive real number and d a positive integer. Show that, for any Dirichlet character χ modulo d,

$$\psi(x, \chi) = \sum_{\substack{1\leq a\leq d\\ (a,d)=1}} \chi(a)\psi(x; d, a).$$

12. Use the Siegel–Walfisz theorem (see Section 7.3 of Chapter 7) and the previous exercise to prove that, for any positive real number x and any non-trivial Dirichlet character χ modulo d with $d \leq (\log x)^N$ for some $N > 0$, we have

$$\psi(x, \chi) \ll x\exp\left(-c\sqrt{\log x}\right)$$

for some positive constant $c = c(N)$.

13. Deduce from Theorem 8.4.1 that, for any $A > 0$ and for any z satisfying $x/(\log x)^A \leq z \leq x$, we have

$$\sum_{d\leq z} \sum_{\substack{1\leq a\leq d\\ (a,d)=1}} \left|\pi(x; d, a) - \frac{\mathrm{li}\,x}{\phi(d)}\right|^2 \leq xz.$$

14. (The Cauchy–Schwarz inequality) Let $f, g : [a, b] \longrightarrow \mathbb{C}$ be differentiable functions, with continuous derivatives. Then

$$\left|\int_a^b f(t)\overline{g(t)}\,dt\right|^2 \leq \left(\int_a^b |f(t)|^2\,dt\right)\left(\int_a^b |g(t)|^2\,dt\right).$$

15. Let z be a positive real number and d a positive integer. Show that

$$\sum_{k \le \frac{z}{d}} \frac{1}{\phi(dk)} \ll \frac{1}{\phi(d)} \log \frac{2z}{d}.$$

16. Let $y > 0, c > 0, T > 0$,

$$\delta(y) := \begin{cases} 0 & \text{if } 0 < y < 1 \\ 1/2 & \text{if } y = 1 \\ 1 & \text{if } y > 1, \end{cases}$$

and

$$I(y, T) := \frac{1}{2\pi i} \int_{c-iT}^{c+iT} \frac{y^s}{s} \, ds.$$

Show that

$$|I(y, T) - \delta(y)| < \begin{cases} y^c \min\left(1, \frac{1}{T|\log y|}\right) & \text{if } y \ne 1 \\ \frac{c}{T} & \text{if } y = 1. \end{cases}$$

17. Use the previous exercise to show that, for any $T > 0, \alpha \in \mathbb{R}, \beta > 0$, we have

$$\int_{-T}^{T} e^{it\alpha} \frac{\sin(t\beta)}{t} \, dt = \begin{cases} \pi + O\left(\frac{1}{T(\beta - |\alpha|)}\right) & \text{if } |\alpha| < \beta \\ \frac{\pi}{2} & \text{if } |\alpha| = \beta \\ O\left(\frac{1}{T(|\alpha| - \beta)}\right) & \text{if } |\alpha| > \beta. \end{cases}$$

18. Let $T > 0$ and x, y be positive integers. Show that

$$\int_{-T}^{T} \min\left(\frac{1}{|t|}, \log(2xy)\right) dt = O\left(\log T + \log(2xy)\right).$$

19. Let $\mathcal{A}$ be a set of Z integers. Let p be a prime and set

$$S\left(\frac{a}{p}\right) := \sum_{n \in \mathcal{A}} e\left(\frac{na}{p}\right),$$

where, as before, $e(t) := \exp(2\pi i t)$. Let $Z(p, h)$ be the number of elements of $\mathcal{A}$ that are $\equiv h \pmod{p}$. Show that

$$\sum_{(a,p)=1} \left| S\left(\frac{a}{p}\right) \right|^2 = p \sum_{h \pmod p} \left(Z(p, h) - \frac{Z}{p}\right)^2.$$

20. With notation as in the previous exercise, let d be an arbitrary positive integer and set

$$S\left(\frac{a}{d}\right) := \sum_{n \in \mathcal{A}} e\left(\frac{na}{d}\right).$$

Let $Z(d, h)$ be the number of elements of $\mathcal{A}$ that are $\equiv h \pmod d$. Let

$$T(d, h) := \sum_{(a,d)=1} S\left(\frac{a}{d}\right) e\left(\frac{-ah}{d}\right).$$

Prove that

$$dZ(d, h) = \sum_{\delta \mid d} T\left(\frac{d}{\delta}, h\right).$$

Deduce that

$$\sum_{(a,d)=1} \left| S\left(\frac{a}{d}\right) \right|^2 = d \sum_{h=1}^{d} \left(\sum_{\delta \mid d} \frac{\mu(\delta)}{\delta} Z\left(\frac{d}{\delta}, h\right) \right)^2.$$

21. Let

$$S\left(\frac{a}{d}\right) := \sum_{n \in \mathcal{A}} a_n e\left(\frac{na}{d}\right),$$

where $\mathcal{A}$ is a finite set of integers. For p prime, let

$$S(p, a) := \sum_{\substack{n \in \mathcal{A} \\ n \equiv a \pmod p}} a_n.$$

and let $\omega(p)$ be the number of $a \pmod p$ for which $S(p, a) = 0$. By observing that

$$S(0) = \sum_{a=1}^{p} S(p, a)$$

and using the Cauchy–Schwarz inequality, show that

$$|S(0)|^2 \left(\frac{\omega(p)}{p - \omega(p)} \right) \le \sum_{(a,p)=1} \left| S\left(\frac{a}{p}\right) \right|^2.$$

22. We keep the notation of the previous exercise. Let d_1, d_2 be coprime positive integers, and let $d = d_1 d_2$. Observing that by the Chinese remainder theorem we have

$$\sum_{(a,d)=1} \left| S\left(\frac{a}{d}\right) \right|^2 = \sum_{(a_1,d_1)=1} \sum_{(a_2,d_2)=1} \left| S\left(\frac{a_1}{d_1} + \frac{a_2}{d_2}\right) \right|^2,$$

deduce that

$$\sum_{(a,d)=1} \left| S\left(\frac{a}{d}\right) \right|^2 \geq |S(0)|^2 J(d_1) J(d_2),$$

where

$$J(d) := \prod_{p|d} \frac{\omega(p)}{p - \omega(p)}.$$

Then deduce that

$$\left| \sum_{n \in \mathcal{A}} a_n \right|^2 \mu^2(d) \prod_{p|d} \frac{\omega(p)}{p - \omega(p)} \leq \sum_{(a,d)=1} \left| S\left(\frac{a}{d}\right) \right|^2.$$

(This gives an alternate method for deriving Theorem 8.2.1.)

9

The Bombieri–Vinogradov theorem

The Bombieri–Vinogradov theorem is considered one of the finest consequences of the large sieve method. Its virtue lies in the fact that in many questions that require the use of the generalized Riemann hypothesis for Dirichlet L-functions, one can apply the Bombieri–Vinogradov theorem. Thus, in situations involving 'sufficiently many' L-functions, the theorem can be regarded as a substitute for the generalized Riemann hypothesis. As we will see below, one application is to the Titchmarsh divisor problem. Another more complicated application can be found in Hooley's monograph [31], where an asymptotic formula is derived for the number of ways of writing a natural number as the sum of a prime number and two squares.

In 1976, Motohashi [43] discovered an induction principle that generalizes the Bombieri–Vinogradov theorem and makes it applicable in a wider context for general arithmetical functions. This perspective was useful in the work of Bombieri *et al.* [2, 3], where the range of summation in the classical Bombieri–Vinogradov theorem is extended beyond the $\sqrt{x}$ barrier. We refer the reader to Section 12 of [1], where a brief discussion of the new ideas can be found.

In this chapter we will derive the classical Bombieri–Vinogradov theorem using a method of Vaughan, as described in [8]. We show that the method has a wide range of applicability by putting it in a general context. We then apply the results to treat the Titchmarsh divisor problem. In the next chapter, we will use the Bombieri–Vinogradov theorem in conjunction with the lower bound sieve method to count the number of primes p such that $p+2$ has a bounded number of prime factors, a result that represents some advance towards the twin prime conjecture.

9.1 A general theorem

The main result of this section is derived as a generalization of a method of Vaughan, which lies at the heart of one proof of the celebrated Bombieri–Vinogradov theorem.

Before stating the general result, let us introduce the class of functions

$$\mathcal{D} := \left\{ D : \mathbb{N} \longrightarrow \mathbb{C} : \sum_{n \leq x} |D(n)|^2 = O\left(x(\log x)^{\alpha}\right) \text{ for some } \alpha > 0 \right\}. \quad (9.1)$$

We have the following basic properties:

Proposition 9.1.1

1. If $D \in \mathcal{D}$ and $\theta \geq 0$, then

$$\sum_{n \leq x} \frac{|D(n)|}{n^{\theta}} \ll x^{1-\theta} (\log x)^{\alpha}$$

for some $\alpha > 0$.
2. If $D_1, D_2 \in \mathcal{D}$, then

$$\sum_{ef \leq x} |D_1(e)D_2(f)| d(ef) \ll x(\log x)^{\beta}$$

for some $\beta > 0$, and

$$\sum_{ef \leq x} |D_1(e)D_2(f)|^2 d(ef) \ll x(\log x)^{\gamma}$$

for some $\gamma > 0$, where for a positive integer e, $d(e)$ denotes the number of the divisors of e.

Proof Exercise. $\square$

Theorem 9.1.2 *Let x, z be positive integers and let*

$$A(s) = \sum_{n \geq 1} \frac{a(n)}{n^s}, \quad B(s) = \sum_{n \geq 1} \frac{b(n)}{n^s}$$

be normalized Dirichlet series (that is, $a(1) = b(1) = 1$) for which we write

$$\frac{A(s)}{B(s)} = \sum_{n \geq 1} \frac{c(n)}{n^s}, \quad \frac{1}{B(s)} = \sum_{n \geq 1} \frac{\tilde{b}(n)}{n^s}$$

for some $c(n), \tilde{b}(n) \in \mathbb{C}$. We assume that all these series are convergent for $\mathrm{Re}(s) > \sigma_0$ for some σ_0, and that they satisfy the following hypotheses:

(H1) $(a(n))_{n \geq 1}$ is an increasing sequence of positive real numbers;
(H2) $b(\cdot), \tilde{b}(\cdot), c(\cdot) \in \mathcal{D}$;
(H3) there exist $0 \leq \theta < 1$ and $0 \leq \gamma < 1$ such that, for any non-trivial Dirichlet character χ modulo d,

$$\sum_{n \leq x} b(n)\chi(n) \ll x^\theta \sqrt{d} \log d + x^\gamma.$$

Then

1. if $z \leq x^{\frac{1-\theta}{3-\theta}}$,

$$\sum_{d \leq z} \frac{d}{\phi(d)} \sum_{\chi}^* \max_{y \leq x} \left| \sum_{n \leq y} c(n)\chi(n) \right| \tag{9.2}$$

$$\ll \left(z^2 x^{\frac{1}{2}} + x + zx^{\frac{5-\theta}{2(3-\theta)}} + z^2 x^{\frac{1-\theta+2\gamma}{3-\theta}} + z^{\frac{5}{2}} x^{\frac{1+\theta}{3-\theta}} a(x) \right) (\log x)^{\alpha'}$$

for some $\alpha' > 0$;

2. if $z > x^{\frac{1-\theta}{3-\theta}}$,

$$\sum_{d \leq z} \frac{d}{\phi(d)} \sum_{\chi}^* \max_{y \leq x} \left| \sum_{n \leq y} c(n)\chi(n) \right| \tag{9.3}$$

$$\ll \left(z^2 x^{\frac{1}{2}} + x + z^{\frac{9-4\theta}{2(3-2\theta)}} x^{\frac{2-\theta}{3-2\theta}} a(x)(\log z) \right.$$

$$\left. + z^{\frac{3-4\theta+3\gamma}{3-2\theta}} x^{\frac{2-2\theta+\gamma}{3-2\theta}} (\log z) \right) (\log x)^{\alpha''}$$

for some $\alpha'' > 0$.

Here, the summation $\sum_{\chi}^*$ is over primitive Dirichlet characters χ modulo d.

Proof We set

$$F(s) := \sum_{n \leq U} \frac{c(n)}{n^s}, \quad G(s) := \sum_{n \leq V} \frac{\tilde{b}(n)}{n^s}$$

for some parameters $U = U(x, z)$ and $V = V(x, z)$, to be chosen later. We think of $F(s)$ as an 'approximation to $A(s)/B(s)$' and of $G(s)$ as 'an approximation to $1/B(s)$', and we observe that we can write

$$\frac{A(s)}{B(s)} = F(s) - B(s)G(s)F(s) + A(s)G(s)$$

$$+ \left(\frac{A(s)}{B(s)} - F(s) \right) (1 - B(s)G(s)). \tag{9.4}$$

In the literature, this is known as **Vaughan's identity** and goes back to ideas of Linnik. By comparing the coefficients of n^{-s} on both sides of (9.4), we deduce that

$$c(n) = a_1(n) + a_2(n) + a_3(n) + a_4(n),$$

where

$$a_1(n) := \begin{cases} c(n) & \text{if} \quad n \leq U \\ 0 & \text{if} \quad n > U, \end{cases} \qquad (9.5)$$

$$a_2(n) := -\sum_{\substack{efg=n \\ f \leq V \\ g \leq U}} b(e)\tilde{b}(f)c(g), \qquad (9.6)$$

$$a_3(n) := \sum_{\substack{ef=n \\ f \leq V}} a(e)\tilde{b}(f), \qquad (9.7)$$

$$a_4(n) := -\sum_{\substack{ef=n \\ e > U \\ f > V}} c(e) \sum_{\substack{gh=f \\ h \leq V}} b(g)\tilde{b}(h). \qquad (9.8)$$

Therefore, for any Dirichlet character χ modulo d we can write

$$\sum_{n \leq y} c(n)\chi(n) = \sum_{1 \leq i \leq 4} \sum_{n \leq y} a_i(n)\chi(n) =: \sum_{1 \leq i \leq 4} S_i(y, \chi). \qquad (9.9)$$

We prove the theorem by estimating each of the sums

$$\sum_{d \leq z} \frac{d}{\phi(d)} {\sum_{\chi}}^* \max_{y \leq x} |S_i(y, \chi)|, \, 1 \leq i \leq 4.$$

The estimate for $S_1(y, \chi)$ Using (9.5) and hypothesis (H2) together with part 1 of Proposition 9.1.1, we obtain

$$|S_1(y, \chi)| = \left| \sum_{\substack{n \leq y \\ n \leq U}} c(n)\chi(n) \right| \ll \sum_{n \leq U} |c(n)| \ll U(\log U)^{\alpha_0}$$

for some $\alpha_0 > 0$. The above bound is independent of y, χ and d, hence, recalling that there are $\phi(d)$ Dirichlet characters modulo d, we obtain

$$\sum_{d \leq z} \frac{d}{\phi(d)} {\sum_{\chi}}^* \max_{y \leq x} |S_1(y, \chi)| \ll z^2 U(\log U)^{\alpha_0}. \qquad (9.10)$$

The estimate for $S_2(y, \chi)$ Using (9.6) we write

$$S_2(y, \chi) = - \sum_{\substack{efg \leq y \\ f \leq V \\ g \leq U}} b(e)\tilde{b}(f)c(g)\chi(efg),$$

which we split into two parts according to whether $fg \leq U$ or $U < fg \leq UV$. The first sum obtained in this way will be denoted by $S_2'(y, \chi)$, and the second sum by $S_2''(y, \chi)$.

For $S_2'(y, \chi)$ we write

$$|S_2'(y, \chi)| \leq \sum_{g \leq U} |c(g)| \sum_{f \leq \min\left(V, \frac{U}{g}\right)} |\tilde{b}(f)| \left| \sum_{e \leq \frac{y}{fg}} b(e)\chi(e) \right|$$

so that we can use hypothesis (H3) to estimate the innermost sum. We get

$$S_2'(y, \chi) \ll y^\theta \sqrt{d}(\log d) \sum_{g \leq U} \frac{|c(g)|}{g^\theta} \sum_{f \leq \min\left(V, \frac{U}{g}\right)} \frac{|\tilde{b}(f)|}{f^\theta}$$

$$+ y^\gamma \sum_{g \leq U} \frac{|c(g)|}{g^\gamma} \sum_{f \leq \min\left(V, \frac{U}{g}\right)} \frac{|\tilde{b}(f)|}{f^\gamma}$$

$$\ll y^\theta \sqrt{d}(\log d) \sum_{g \leq U} \frac{|c(g)|}{g^\theta} \sum_{f \leq \frac{U}{g}} \frac{|\tilde{b}(f)|}{f^\theta}$$

$$+ y^\gamma \sum_{g \leq U} \frac{|c(g)|}{g^\gamma} \sum_{f \leq \frac{U}{g}} \frac{|\tilde{b}(f)|}{f^\gamma}.$$

Then, by using (H2) and part 1 of Proposition 9.1.1, we obtain

$$S_2'(y, \chi) \ll y^\theta \sqrt{d}(\log d) \sum_{g \leq U} \frac{|c(g)|}{g^\theta} \left(\frac{U}{g}\right)^{1-\theta} (\log U)^{\alpha_1}$$

$$+ y^\gamma \sum_{g \leq U} \frac{|c(g)|}{g^\gamma} \left(\frac{U}{g}\right)^{1-\gamma} (\log U)^{\alpha_2}$$

$$\ll y^\theta \sqrt{d}(\log d)U^{1-\theta}(\log U)^{\alpha_3} + y^\gamma U^{1-\gamma}(\log U)^{\alpha_4}$$

for some $\alpha_1, \alpha_2, \alpha_3, \alpha_4 > 0$. This implies that

$$\sum_{d \leq z} \frac{d}{\phi(d)} \sum_\chi^* \max_{y \leq x} |S_2'(y, \chi)|$$

$$\ll x^\theta z^{\frac{5}{2}}(\log z)U^{1-\theta}(\log U)^{\alpha_3} + x^\gamma z^2(\log z)U^{1-\gamma}(\log U)^{\alpha_4}. \quad (9.11)$$

For $S_2''(y, \chi)$ we write

$$\sum_{d \leq z} \frac{d}{\phi(d)} \sideset{}{^*}\sum_{\chi} \max_{y \leq x} |S_2''(y, \chi)|$$

$$= \sum_{d \leq z} \frac{d}{\phi(d)} \sideset{}{^*}\sum_{\chi} \max_{y \leq x} \left| \sum_{\substack{eh \leq y \\ U < h \leq UV}} b(e) \left(\sum_{\substack{f \leq V, g \leq U \\ fg = h}} \tilde{b}(f)c(g) \right) \chi(eh) \right|,$$

which suggests that we use the second modified large sieve inequality discussed in Chapter 8. However, we observe that by applying this inequality directly to the pair of sequences of complex numbers

$$\left(b(e) \right)_{\frac{x}{UV} < e \leq \frac{x}{U}} \quad \text{and} \quad \left(\tilde{b}(f)c(g) \right)_{fg = h, U < h \leq UV},$$

we will overestimate our expression. Instead, we should divide the interval $(U, UV]$ into dyadic intervals $(2^k, 2^{k+1}]$ with $[\log_2 U] < k < [\log_2 UV]$ and apply the modified large sieve inequality to each of the pair of sequences

$$\left(b(e) \right)_{\frac{x}{2^{k+1}} < e \leq \frac{x}{2^k}} \quad \text{and} \quad \left(\tilde{b}(f)c(g) \right)_{fg = h, 2^k < h \leq 2^{k+1}}.$$

For each $[\log_2 U] < k < [\log_2 UV]$ we obtain that

$$\sum_{d \leq z} \frac{d}{\phi(d)} \sideset{}{^*}\sum_{\chi} \max_{y \leq x} \left| \sum_{\substack{eh \leq y \\ 2^k < h \leq 2^{k+1}}} b(e) \left(\sum_{\substack{f \leq V, g \leq U \\ fg = h}} \tilde{b}(f)c(g) \right) \chi(eh) \right|$$

$$\ll \left(z^2 + \frac{x}{2^k} \right)^{1/2} \left(z^2 + 2^k \right)^{1/2} \left(\sideset{}{'}\sum_e \right)^{1/2} \left(\sideset{}{'}\sum_h \right)^{1/2} (\log x), \quad (9.12)$$

where

$$\sideset{}{'}\sum_e := \sum_{e \leq \frac{x}{2^k}} |b(e)|^2, \quad (9.13)$$

$$\sideset{}{'}\sum_h := \sum_{2^k < h \leq 2^{k+1}} \left| \sum_{\substack{f \leq V, g \leq U \\ fg = h}} \tilde{b}(f)c(g) \right|^2. \quad (9.14)$$

By (H2) we get

$$\sum_{e \leq \frac{x}{2^k}} |b(e)|^2 \ll \frac{x}{2^k} \left(\log \frac{x}{2^k} \right)^{\alpha_5} \quad (9.15)$$

for some $\alpha_5 > 0$, and by the Cauchy–Schwarz inequality together with (H2) and part 2 of Proposition 9.1.1, we get

$$\sum_{2^k < h \leq 2^{k+1}} \left| \sum_{\substack{f \leq V, g \leq U \\ fg = h}} \tilde{b}(f) c(g) \right|^2 \ll \sum_{2^k < h \leq 2^{k+1}} d(h) \sum_{f | h} |\tilde{b}(f)|^2 \left| c\left(\frac{h}{f}\right) \right|^2$$

$$\ll 2^k \left(\log 2^k \right)^{\alpha_6} \tag{9.16}$$

for some $\alpha_6 > 0$. Then we plug (9.15) and (9.16) into (9.13), (9.14), (9.12), to obtain

$$\sum_{d \leq z} \frac{d}{\phi(d)} {\sum_{\chi}}^* \max_{y \leq x} \left| \sum_{\substack{eh \leq y \\ 2^k < h \leq 2^{k+1}}} b(e) \left(\sum_{\substack{f \leq V, g \leq U \\ fg = h}} \tilde{b}(f) c(g) \right) \chi(eh) \right|$$

$$\ll \left(z^2 + \frac{zx^{1/2}}{2^{k/2}} + z2^{k/2} + x^{1/2} \right) x^{\frac{1}{2}} (\log x)^{\frac{\alpha_5}{2}+1} \left(\log 2^k \right)^{\frac{\alpha_6}{2}}.$$

We sum over all k to finally get

$$\sum_{d \leq z} \frac{d}{\phi(d)} {\sum_{\chi}}^* \max_{y \leq x} |S_2''(y, \chi)| \tag{9.17}$$

$$\ll \left(z^2 + \frac{zx^{1/2}}{U^{1/2}} + z(UV)^{1/2} + x^{1/2} \right) x^{\frac{1}{2}} (\log x)^{\frac{\alpha_5}{2}+1} (\log UV)^{\frac{\alpha_6}{2}+1}.$$

The estimate for $S_3(y, \chi)$ We define a step function $\mathcal{A} : \mathbb{R} \longrightarrow \mathbb{R}$ by $\mathcal{A}(t) = a(1)$ if $t \leq 1$, $\mathcal{A}(t) = a(2) - a(1)$ if $1 < t \leq 2$, and, generally, $\mathcal{A}(t) = a(n) - a(n-1)$ if $n - 1 < t \leq n$. Then we observe that $a(n) = \int_0^n \mathcal{A}(t)\, dt$ and that $\mathcal{A}(\cdot)$ is positive, since the sequence $(a(n))_{n \geq 1}$ is increasing. Using (9.7) (and assuming, without loss of generality, that y is an integer) we write

$$|S_3(y, \chi)| = \left| \sum_{f \leq V} \tilde{b}(f) \chi(f) \sum_{e \leq \frac{y}{f}} a(e) \chi(e) \right|$$

$$= \left| \sum_{f \leq V} \tilde{b}(f) \chi(f) \sum_{e \leq \frac{y}{f}} \chi(e) \int_0^e \mathcal{A}(t)\, dt \right|$$

$$= \left| \int_0^y \mathcal{A}(t) \sum_{f \leq V} \tilde{b}(f) \chi(f) \sum_{t \leq e \leq \frac{y}{f}} \chi(e) \, dt \right|$$

$$\leq \int_0^y |\mathcal{A}(t)| \sum_{f \leq V} |\tilde{b}(f)| \left| \sum_{t \leq e \leq \frac{y}{f}} \chi(e) \right| dt.$$

We remark that we can use the Pólya–Vinogradov inequality (8.14) to estimate the inner sum. We obtain

$$S_3(y, \chi) \ll \sqrt{d}(\log d) \int_0^y |\mathcal{A}(t)| \, dt \sum_{f \leq V} |\tilde{b}(f)| \ll \sqrt{d}(\log d) V(\log V)^{\alpha_7} a(y)$$

for some $\alpha_7 > 0$. Therefore, by observing that $|\mathcal{A}(t)| = \mathcal{A}(t)$ and by using (H1), we further obtain

$$\sum_{d \leq z} \frac{d}{\phi(d)} \sum_{\chi}^* \max_{y \leq x} |S_3(y, \chi)| \ll z^{\frac{5}{2}} (\log z) V(\log V)^{\alpha_7} \max_{y \leq x} |a(y)|$$

$$= z^{\frac{5}{2}} (\log z) V(\log V)^{\alpha_7} a(x). \qquad (9.18)$$

The estimate for $S_4(y, \chi)$ By using (9.8) we write

$$\sum_{d \leq z} \frac{d}{\phi(d)} \sum_{\chi}^* \max_{y \leq x} |S_4(y, \chi)|$$

$$= \sum_{d \leq z} \frac{d}{\phi(d)} \sum_{\chi}^* \max_{y \leq x} \left| \sum_{\substack{ef \leq y \\ e > U \\ f > V}} c(e) \left(\sum_{\substack{gh = f \\ h \leq V}} b(g) \tilde{b}(h) \right) \chi(ef) \right|,$$

which suggests that we apply the second modified large sieve inequality to the pair of sequences of complex numbers

$$(c(e))_{U < e < \frac{x}{V}} \quad \text{and} \quad \left(b(g) \tilde{b}(h) \right)_{gh = f, h \leq V, V < f < \frac{x}{U}}.$$

However, in order to obtain optimal results, we proceed as when handling the sum $S_2''(y, \chi)$. To be precise, we divide the interval $(U, y/V]$ into dyadic intervals $(2^k, 2^{k+1}]$ with $[\log_2 U] < k \leq [\log_2 y/V]$ and we apply the modified large sieve inequality to each of the pair of sequences

$$(c(e))_{2^k < e \leq 2^{k+1}} \quad \text{and} \quad \left(b(g) \tilde{b}(h) \right)_{gh = f, h \leq V, \max(V, \frac{y}{2^{k+1}}) < f < \frac{y}{2^k}}.$$

For each $[\log_2 U] < k \le \left[\log_2 \frac{y}{V}\right]$ we obtain

$$
\sum_{d \le z} \frac{d}{\phi(d)} \sum_{\chi}^{*} \max_{y \le x} \left| \sum_{\substack{ef \le y \\ e > U \\ f > V \\ 2^k < e \le 2^{k+1}}} c(e) \left(\sum_{\substack{gh = f \\ h \le V}} b(g)\tilde{b}(h) \right) \chi(ef) \right|
$$

$$
\ll \left(z^2 + 2^k\right)^{1/2} \left(z^2 + \frac{x}{2^k}\right)^{1/2} \left(\sideset{}{''}\sum_{e}\right)^{1/2} \left(\sideset{}{''}\sum_{f}\right)^{1/2} \log x, \qquad (9.19)
$$

where

$$
\sideset{}{''}\sum_{e} := \sum_{2^k < e \le 2^{k+1}} |c(e)|^2, \qquad (9.20)
$$

$$
\sideset{}{''}\sum_{f} := \sum_{V < f \le \frac{x}{2^k}} \left| \sum_{\substack{gh = f \\ h \le V}} b(g)\tilde{b}(h) \right|^2. \qquad (9.21)
$$

We observe that by (H2) we get

$$
\sum_{2^k < e \le 2^{k+1}} |c(e)|^2 \ll 2^k \left(\log 2^k\right)^{\alpha_8} \qquad (9.22)
$$

for some $\alpha_8 > 0$, and by the Cauchy–Schwarz inequality together with (H2) and part 2 of Proposition 9.1.1, we get

$$
\sum_{V < f \le \frac{x}{2^k}} \left| \sum_{\substack{gh = f \\ h \le V}} b(g)\tilde{b}(h) \right|^2 \ll \sum_{V < f \le \frac{x}{2^k}} d(f) \sum_{h|f} |\tilde{b}(h)|^2 \left| b\left(\frac{f}{h}\right) \right|^2
$$

$$
\ll \frac{x}{2^k} \left(\log \frac{x}{2^k}\right)^{\alpha_9} \qquad (9.23)
$$

for some $\alpha_9 > 0$. We plug (9.22) and (9.23) into (9.20), (9.21), (9.19), and obtain

$$
\sum_{d \le z} \frac{d}{\phi(d)} \sum_{\chi}^{*} \max_{y \le x} \left| \sum_{\substack{ef \le y \\ e > U \\ f > V \\ 2^k < e \le 2^{k+1}}} c(e) \left(\sum_{\substack{gh = f \\ h \le V}} b(g)\tilde{b}(h) \right) \chi(ef) \right|
$$

$$
\ll \left(z^2 + \frac{zx^{1/2}}{2^{k/2}} + z 2^{k/2} + x^{1/2}\right) x^{\frac{1}{2}} (\log x)^{\frac{\alpha_9}{2}+1} \left(\log 2^k\right)^{\frac{\alpha_8}{2}}.
$$

Finally we sum over all k and deduce that

$$\sum_{d \leq z} \frac{d}{\phi(d)} \sum_{\chi}^{*} \max_{y \leq x} |S_4(y, \chi)|$$

$$\ll \left(z^2 + \frac{zx^{1/2}}{U^{1/2}} + \frac{zx^{1/2}}{V^{1/2}} + x^{1/2} \right) x^{\frac{1}{2}} (\log x)^{\frac{\alpha_8 + \alpha_9}{2} + 2}. \qquad (9.24)$$

We have completed estimating all the four sums $S_i(y, \chi)$, $1 \leq i \leq 4$ (see (9.10), (9.11), (9.17), (9.18) and (9.24)). Putting these estimates together gives that

$$\sum_{d \leq z} \frac{d}{\phi(d)} \sum_{\chi}^{*} \max_{y \leq x} \left| \sum_{n \leq y} c(n) \chi(n) \right|$$

$$\ll z^2 U (\log U)^{\alpha_0}$$

$$\quad + x^{\theta} z^{5/2} (\log z) U^{1-\theta} (\log U)^{\alpha_3} + x^{\gamma} z^2 (\log z) U^{1-\gamma} (\log U)^{\alpha_4}$$

$$\quad + \left(z^2 + \frac{zx^{1/2}}{U^{1/2}} + z(UV)^{1/2} + x^{1/2} \right) x^{\frac{1}{2}} (\log x)^{\frac{\alpha_5}{2} + 1} (\log UV)^{\frac{\alpha_6}{2} + 1}$$

$$\quad + z^{5/2} (\log z) V (\log V)^{\alpha_7} a(x)$$

$$\quad + \left(z^2 + \frac{zx^{1/2}}{U^{1/2}} + \frac{zx^{1/2}}{V^{1/2}} + x^{1/2} \right) x^{\frac{1}{2}} (\log x)^{\frac{\alpha_8 + \alpha_9}{2} + 2}. \qquad (9.25)$$

It remains to choose the parameters U and V appropriately. We look for U and V such that $z^{5/2} x^{\theta} U^{1-\theta} = z^{5/2} V$, that is, such that

$$V = x^{\theta} U^{1-\theta}.$$

Now we analyse the expression

$$E(x, z, U) := \frac{zx}{U^{1/2}} + z^{5/2} x^{\theta} U^{1-\theta} + zx^{\frac{1}{2}} (UV)^{1/2}$$

$$= \frac{zx}{U^{1/2}} + zU \left(z^{\frac{3}{2}} x^{\theta} U^{-\theta} + x^{\frac{1+\theta}{2}} U^{-\frac{\theta}{2}} \right).$$

If

$$z^{\frac{3}{2}} x^{\theta} U^{-\theta} \leq x^{\frac{1+\theta}{2}} U^{-\frac{\theta}{2}},$$

that is, if

$$z \leq x^{\frac{1-\theta}{3}} U^{\frac{\theta}{3}}, \qquad (9.26)$$

then

$$E(x, z, U) \ll \frac{zx}{U^{1/2}} + zx^{\frac{1+\theta}{2}} U^{\frac{2-\theta}{2}}.$$

We choose U such that

$$\frac{zx}{U^{1/2}} = zx^{\frac{1+\theta}{2}} U^{\frac{2-\theta}{2}},$$

that is,

$$U := x^{\frac{1-\theta}{3-\theta}}.$$

Going back to (9.26), we note that our choice of U implies

$$z \leq x^{\frac{1-\theta}{3-\theta}}.$$

We obtain

$$\sum_{d \leq z} \frac{d}{\phi(d)} \sideset{}{^*}\sum_{\chi} \max_{y \leq x} \left| \sum_{n \leq y} c(n)\chi(n) \right|$$

$$\ll \left(z^2 x^{\frac{1}{2}} + x + zx^{\frac{5-\theta}{2(3-\theta)}} + z^2 x^{\frac{1-\theta+2\gamma}{3-\theta}} + z^{\frac{5}{2}} x^{\frac{1+\theta}{3-\theta}} a(x) \right) (\log x)^{\alpha'}$$

for some $\alpha' > 0$.

If

$$z^{\frac{3}{2}} x^{\theta} U^{-\theta} > x^{\frac{1+\theta}{2}} U^{-\frac{\theta}{2}},$$

then we proceed as above and choose U such that

$$\frac{zx}{U^{1/2}} = z^{5/2} x^{\theta} U^{1-\theta}.$$

Thus

$$U := \frac{x^{\frac{2(1-\theta)}{3-2\theta}}}{z^{\frac{3}{3-2\theta}}}.$$

With this choice of U we get that

$$z > x^{\frac{1-\theta}{3-\theta}},$$

and that

$$\sum_{d \leq z} \frac{d}{\phi(d)} \sideset{}{^*}\sum_{\chi} \max_{y \leq x} \left| \sum_{n \leq y} c(n)\chi(n) \right|$$

$$\ll \left(z^2 x^{\frac{1}{2}} + x + z^{\frac{9-4\theta}{2(3-2\theta)}} x^{\frac{2-\theta}{3-2\theta}} a(x)(\log z) + z^{\frac{3-4\theta+3\gamma}{3-2\theta}} x^{\frac{2-2\theta+\gamma}{3-2\theta}} (\log z) \right) (\log x)^{\alpha''}$$

for some $\alpha'' > 0$.

This completes the proof of the theorem. $\square$

9.2 The Bombieri–Vinogradov theorem

In this section we return to the study of the error term that occurs in the asymptotic formula for $\pi(x; d, a)$, where a, d are coprime integers and $d \leq x$. We have already mentioned in Chapter 7 that if we assume a generalized Riemann hypothesis, then

$$\pi(x; d, a) = \frac{\mathrm{li}\, x}{\phi(d)} + O\left(x^{1/2} \log(dx)\right).$$

We are now ready to prove that, on 'average' and under no hypothesis, the error term has indeed size $O_d(x^{1/2} \log x)$. More precisely, we shall be proving:

Theorem 9.2.1 *(The Bombieri–Vinogradov theorem)*
For any $A > 0$ there exists $B = B(A) > 0$ such that

$$\sum_{d \leq \frac{x^{1/2}}{(\log x)^B}} \max_{y \leq x} \max_{(a,d)=1} \left| \pi(y; d, a) - \frac{\mathrm{li}\, y}{\phi(d)} \right| \ll \frac{x}{(\log x)^A}. \tag{9.27}$$

The Bombieri–Vinogradov theorem was first obtained, independently, by E. Bombieri and A. I. Vinogradov in 1965. Their proofs were based on an application of the large sieve for estimating the number of zeroes of certain L-functions in various rectangles. A different proof of the theorem was obtained by Gallagher in 1968. Yet another proof was obtained by Vaughan in 1975, and it is his approach that we will be following in our presentation below (see [8, p.16]).

There are two main ingredients needed in Vaughan's proof of the Bombieri–Vinogradov theorem. The first one is a particular case of the general theorem that we discussed in Section 9.1. We state it below. The second one is the result due to Siegel and Walfisz, stated in Section 7.3 of Chapter 7, which provides unconditional estimates for $\pi(x; d, a) - \mathrm{li}\, x/\phi(d)$ as long as d runs in small ranges with respect to x.

Theorem 9.2.2 *(Vaughan)*
Let x and z be arbitrary positive integers. Then

$$\sum_{d \leq z} \frac{d}{\phi(d)} \sum_{\chi}^{*} \max_{y \leq x} \left| \sum_{n \leq y} \Lambda(n)\chi(n) \right|$$

$$\ll \left(z^2 x^{1/2} + x + z x^{5/6}\right) (\log z)(\log x)^\alpha \tag{9.28}$$

for some $\alpha > 0$, where the summation $\sum_{\chi}^{*}$ is over primitive characters χ modulo d and where $\Lambda(\cdot)$ denotes the von Mangoldt function.

Proof　We use the notation introduced in Theorem 9.1.2 and set

$$A(s) := -\zeta'(s) = \sum_{n \geq 1} \frac{\log n}{n^s}, \quad B(s) := \zeta(s) = \sum_{n \geq 1} \frac{1}{n^s}.$$

Thus, for any $n \geq 1$,

$$a(n) = \log n, \, b(n) = 1, \, c(n) = \Lambda(n), \, \tilde{b}(n) = \mu(n),$$

and hypotheses (H1) and (H2) of Theorem 9.1.2 are clearly satisfied. By using the Pólya–Vinogradov inequality (8.14) we see that, for any d and any non-trivial Dirichlet character χ modulo d,

$$\sum_{n \leq x} b(n)\chi(n) = \sum_{n \leq x} \chi(n) \ll \sqrt{d} \log d.$$

Hence hypothesis (H3) is also satisfied, with $\theta = 0 = \gamma$. We obtain that if $z \leq x^{1/3}$, then

$$\sum_{d \leq z} \frac{d}{\phi(d)} {\sum_{\chi}}^* \max_{y \leq x} \left| \sum_{n \leq y} \Lambda(n)\chi(n) \right|$$

$$\ll \left(z^2 x^{1/2} + x + z x^{5/6} + z^{5/2} x^{1/3} \right) (\log z)(\log x)^{\alpha'}$$

for some $\alpha' > 0$, and if $z > x^{1/3}$, then

$$\sum_{d \leq z} \frac{d}{\phi(d)} {\sum_{\chi}}^* \max_{y \leq x} \left| \sum_{n \leq y} \Lambda(n)\chi(n) \right|$$

$$\ll \left(z^2 x^{1/2} + x + z^{3/2} x^{2/3} \right) (\log z)(\log x)^{\alpha''}$$

for some $\alpha'' > 0$. Combining the two estimates gives us the desired inequality. □

Corollary 9.2.3　*Let x, z, D be positive integers such that $z > D$. Then*

$$\sum_{D < d \leq z} \frac{d}{\phi(d)} {\sum_{\chi}}^* \max_{y \leq x} \left| \sum_{n \leq y} \Lambda(n)\chi(n) \right|$$

$$\ll \left(z x^{1/2} + \frac{x}{z} + \frac{x}{D} + x^{5/6} \log z \right) (\log z)(\log x)^{\alpha} \quad (9.29)$$

for some $\alpha > 0$, where, again, the summation ${\sum_{\chi}}^*$ is over primitive Dirichlet characters χ modulo d.

Proof　This is an immediate consequence of Theorem 9.2.2 and of the method of partial summation. We leave it to the reader as an exercise. □

Proof of the Bombieri–Vinogradov theorem We will follow very closely the proof of Theorem 8.4.1 of Chapter 8. To obtain the stronger estimate (9.27) we will use, instead of the modified large sieve inequality, the technical result given by Corollary 9.2.3. The details follow.

First we remark that proving (9.27) is equivalent to proving that for any $A > 0$ there exists $B = B(A) > 0$ such that

$$\sum_{d \leq \frac{x^{1/2}}{(\log x)^B}} \max_{y \leq x} \max_{(a,d)=1} \left| \psi(y; d, a) - \frac{y}{\phi(d)} \right| \ll \frac{x}{(\log x)^A},$$

where $\psi(y; d, a)$ is defined in (8.24).

Let A, y, x be positive real numbers such that $y \leq x$, and let $d \leq y$. We observe that

$$\max_{(a,d)=1} \left| \psi(y; d, a) - \frac{y}{\phi(d)} \right| \leq \frac{1}{\phi(d)} \sum_{\substack{\chi \pmod d \\ \chi \neq \chi_0}} |\psi(y, \chi)| + \frac{\psi(y, \chi_0) - y}{\phi(d)}. \quad (9.30)$$

Let the character $\chi \neq \chi_0$ modulo d be induced by some primitive character χ_1 modulo d_1. We saw in (8.28) that

$$\psi(y, \chi_1) - \psi(y, \chi) \ll (\log y)(\log d).$$

Using this in (9.30) gives us that

$$\max_{(a,d)=1} \left| \psi(y; d, a) - \frac{y}{\phi(d)} \right| \ll \frac{1}{\phi(d)} \sum_{\substack{\chi \pmod d \\ \chi \neq \chi_0}} |\psi(y, \chi_1)|$$

$$+ \sum_{d \leq z} \frac{1}{\phi(d)} \max_{y \leq x} |\psi(y) - y| + (\log y)(\log d),$$

and, further, that

$$\sum_{d \leq z} \max_{y \leq x} \max_{(a,d)=1} \left| \psi(y; d, a) - \frac{y}{\phi(d)} \right| \ll \sum_{d \leq z} \frac{1}{\phi(d)} \sum_{\substack{\chi \pmod d \\ \chi \neq \chi_0}} \max_{y \leq x} |\psi(y, \chi_1)|$$

$$+ \sum_{d \leq z} \frac{1}{\phi(d)} \max_{y \leq x} |\psi(y) - y| + z(\log z)(\log x),$$

where $z = z(x)$ is a positive real number, depending on x, to be specified soon. For the second term above we use the prime number theorem in the form (8.29) given in Chapter 8, obtaining that

$$\sum_{d \leq z} \frac{1}{\phi(d)} \max_{y \leq x} |\psi(y) - y| \ll \frac{x \log z}{(\log x)^{A+1}}. \quad (9.31)$$

It remains to estimate the first term.

We write each modulus d as $d = d_1 k$ for some positive integer k and observe that

$$\sum_{d \leq z} \frac{1}{\phi(d)} \sum_{\substack{\chi \pmod d \\ \chi \neq \chi_0}} \max_{y \leq x} |\psi(y, \chi_1)| = \sum_{d_1 \leq z} \sum_{k \leq \frac{z}{d_1}} \frac{1}{\phi(d_1 k)} \sum_{\chi_1 \pmod{d_1}} \max_{y \leq x} |\psi(y, \chi_1)|$$

$$\ll \sum_{d \leq z} \frac{1}{\phi(d)} \log\left(\frac{2z}{d}\right) \sideset{}{^*}\sum_{\chi} |\psi(y, \chi)|,$$

where we have also used the estimate

$$\sum_{k \leq \frac{z}{d}} \frac{1}{\phi(dk)} \ll \frac{1}{\phi(d)} \log \frac{2z}{d}$$

and where the summation $\sum_{\chi}^*$ is over primitive characters χ modulo d. In order to complete the proof of the theorem we need to choose z appropriately (i.e. of the form $z = x^{1/2}/(\log x)^B$ for some positive constant $B = B(A)$) and to show that

$$\sum_{d \leq z} \frac{1}{\phi(d)} \sideset{}{^*}\sum_{\chi} \max_{y \leq x} |\psi(y, \chi)| \ll \frac{x}{(\log x)^A}. \tag{9.32}$$

We recall that by the Siegel–Walfisz theorem, namely by Exercise 12 of Chapter 8, there exists $B = B(A) > 0$ such that, if $d \leq (\log x)^B$ and $\chi \neq \chi_0$ is a character modulo d, then

$$\psi(y, \chi) \ll \frac{x}{(\log x)^{A+1}}.$$

Thus

$$\sum_{d \leq (\log x)^B} \frac{1}{\phi(d)} \sideset{}{^*}\sum_{\chi} \max_{y \leq x} |\psi(y, \chi)| \ll \frac{x}{(\log x)^A}. \tag{9.33}$$

Now we choose

$$z := \frac{x^{1/2}}{(\log x)^B}$$

and apply Corollary 9.2.3 and partial summation to obtain

$$\sum_{(\log x)^B < d \leq z} \frac{1}{\phi(d)} \sideset{}{^*}\sum_{\chi} \max_{y \leq x} |\psi(y, \chi)| \ll \frac{x}{(\log x)^A}. \tag{9.34}$$

By combining (9.31), (9.33) and (9.34), the proof of the theorem is completed. $\square$

The Bombieri–Vinogradov theorem has been extended by Bombieri *et al.* [3] in the following way. Let $a \neq 0$ and $x \geq y \geq 3$. Then

$$\sum_{\substack{d \leq x^{\frac{1}{2}} y^{\frac{1}{2}} \\ (d,a)=1}} \left| \psi(x; d, a) - \frac{x}{\phi(d)} \right| \ll x \left(\frac{\log y}{\log x} \right)^2 (\log \log x)^B.$$

Here B is an absolute constant and the constant implied by the $\ll$ symbol depends only on a.

This means that in most applications involving the Bombieri–Vinogradov theorem, one can take

$$d \leq x^{\frac{1}{2}} \exp \left(\frac{\log x}{(\log \log x)^B} \right)$$

instead of $d \leq x^{\frac{1}{2}} (\log x)^{-B}$. In particular, this applies to the Titchmarsh divisor problem (to be discussed in the following section) and we can deduce the existence of positive constants c and c_1 such that

$$\sum_{p \leq x} d(p+a) = cx + c_1 \frac{x}{\log x} + O\left(\frac{x(\log \log x)^B}{(\log x)^2} \right).$$

The problem of proving results of the form

$$\sum_{d \leq x^\theta} \max_{(a,d)=1} \left| \psi(x; d, a) - \frac{x}{\phi(d)} \right| \ll \frac{x}{(\log x)^A}$$

for any $A > 0$ and $\theta > 1/2$ is an exceedingly difficult one. A celebrated **conjecture of Elliott and Halberstam** [13] predicts that the above inequality is true for any $\theta < 1$.

If one discards the absolute value and fixes a, then some significant results have been obtained in [2]. Without going into too much detail, we mention that for so-called 'well-factorable functions $\lambda(d)$ of level z' we have, for any $\varepsilon > 0$ and $z := x^{\frac{4}{7}-\varepsilon}$,

$$\sum_{\substack{(d,a)=1 \\ d \leq z}} \lambda(d) \left(\psi(x; d, a) - \frac{x}{\phi(d)} \right) \ll \frac{x}{(\log x)^A}$$

for any $A > 0$.

This result has numerous applications. For example, in the case of the Titchmarsh divisor prolem one can show, for any $A > 0$,

$$\sum_{p \leq x} d(p+a) = cx + c_1 \mathrm{li}\, x + O\left(\frac{x}{(\log x)^A} \right).$$

for positive constants c, c_1. Another aplication is to the Artin primitive root conjecture [44], as well as to the theory of elliptic curves.

The proofs of these improvements depend upon estimates for averages of Kloosterman sums and these, in turn, are derived from the fundamental work of Deshouillers and Iwaniec [9] relating spectral theory to sieve theory.

In conclusion, we mention an important paper of Fouvry [18] that signals a conjectural approach to the Elliott–Halberstam conjecture.

9.3 The Titchmarsh divisor problem

Let a be a fixed integer. We recall that in Chapter 3 we considered the question of determining the asymptotic behaviour of the function

$$\sum_{p \leq x} \nu(p+a)$$

for $a = -1$. Now let us consider the more complex question of determining the asymptotic behaviour of

$$\sum_{p \leq x} d(p+a),$$

where $d(\cdot)$ is the divisor function. This is known in the literature as the **Titchmarsh divisor problem**. It was first studied by Titchmarsh in 1930 [69] and is related to a famous **conjecture of Hardy and Littlewood**, formulated in 1922 and asserting that every sufficiently large integer can be represented as the sum of a prime and two squares. Already in 1930, Titchmarsh showed that

$$\sum_{p \leq x} d(p+a) = O(x),$$

and that, under a generalized Riemann hypothesis, an explicit asymptotic formula for $\sum_{p \leq x} d(p+a)$ (see (9.35) below) also holds.

In 1923, Hardy and Littlewood suggested that their aforementioned conjecture was true for almost all integers if one assumes a generalized Riemann hypothesis, and in 1928 Stanley showed that this was indeed true. Later on, the hypothesis of her results was gradually weakened by S. Chowla (1935), A. Walfisz (1935), T. Estermann (1936), H. Halberstam (1951), and C. Hooley (1957), without being completely removed. Only in 1960 Linnik obtained an unconditional proof for the Hardy–Littlewood conjecture. His method, now known as 'the dispersion method', can also be used to provide an unconditional proof for the Titchmarsh divisor problem.

Simpler proofs for both the Hardy–Littlewood conjecture and the Titchmarsh divisor problem were obtained in 1966 by P. D. T. A. Elliott and H. Halberstam, and in 1965 by G. Rodriquez, by combining Hooley's 1957 approach (see [31]) with the Bombieri–Vinogradov Theorem.

Our goal in this section is to describe a simple proof of the Titchmarsh divisor problem. More precisely, we are showing:

Theorem 9.3.1 *Let a be a fixed integer. Then there exists a positive constant c such that*

$$\sum_{p \leq x} d(p+a) = cx + O\left(\frac{x \log \log x}{\log x}\right) \qquad (9.35)$$

Proof First let us observe that for any positive integer n,

$$d(n) = 2 \sum_{\substack{d \mid n \\ d \leq \sqrt{n}}} 1 - \delta(n),$$

where

$$\delta(n) := \begin{cases} 1 & \text{if} \quad n \text{ is a square} \\ 0 & \text{otherwise.} \end{cases}$$

Thus

$$\sum_{p \leq x} d(p+a) = 2 \sum_{p \leq x} \sum_{\substack{d \mid p+a \\ d \leq \sqrt{x}}} 1 - \sum_{p \leq x} \delta(p+a)$$

$$= 2 \sum_{d \leq \sqrt{x}} \pi(x; d, -a) + O\left(\sqrt{x}\right). \qquad (9.36)$$

We recall that the Bombieri–Vinogradov theorem allows us to control the error terms in the asymptotic formulae for $\pi(x; d, -a)$, as long as $d \leq \sqrt{x}(\log x)^{-B}$ for some positive constant B (to be specified later). This suggests that we split the summation on the right-hand side of (9.36) into two parts:

$$\sum_{d \leq \sqrt{x}} \pi(x; d, -a) = \sum_{d \leq \frac{\sqrt{x}}{(\log x)^B}} \pi(x; d, -a) + \sum_{\frac{\sqrt{x}}{(\log x)^B} < d \leq \sqrt{x}} \pi(x; d, -a). \qquad (9.37)$$

For the first sum in (9.37) we write

$$\sum_{d \le \frac{\sqrt{x}}{(\log x)^B}} \pi(x; d, -a) = \sum_{d \le \frac{\sqrt{x}}{(\log x)^B}} \left(\pi(x; d, -a) - \frac{\operatorname{li} x}{\phi(d)} \right) + \sum_{d \le \frac{\sqrt{x}}{(\log x)^B}} \frac{\operatorname{li} x}{\phi(d)}$$

and use the Bombieri–Vinogradov theorem to obtain an upper bound of

$$\operatorname{li} x \sum_{d \le \frac{\sqrt{x}}{(\log x)^B}} \frac{1}{\phi(d)} + O\left(\frac{x}{(\log x)^A} \right).$$

for any arbitrary $A > 0$ and some $B = B(A)$. Now we remark (see Exercise 4 of Chapter 6) that there exists a positive constant c_0 such that, for any x,

$$\sum_{d \le x} \frac{1}{\phi(d)} = c_0 \log x + O(1). \tag{9.38}$$

Therefore

$$\sum_{d \le \frac{\sqrt{x}}{(\log x)^B}} \pi(x; d, -a) = \frac{c_0}{2} x + O\left(\frac{x \log \log x}{\log x} \right). \tag{9.39}$$

For the second sum in (9.37) we use the Brun–Titchmarsh theorem and, again, (9.38). We obtain

$$\sum_{\frac{\sqrt{x}}{(\log x)^B} < d \le \sqrt{x}} \pi(x; d, -a) \ll \frac{x}{\log x} \sum_{\frac{\sqrt{x}}{(\log x)^B} < d \le \sqrt{x}} \frac{1}{\phi(d)} \ll \frac{x \log \log x}{\log x}. \tag{9.40}$$

The proof of the theorem is completed by combining (9.36), (9.37), (9.39) and (9.40), and by setting $c := c_0/2$. □

9.4 Exercises

1. Use partial summation and the Cauchy–Schwarz inequality to prove Proposition 9.1.1.
2. Let $n, k \ge 1$ and let $d_k(n)$ denote the number of ways of representing n as a product of k positive integers. Observe that $d_2(n) = d(n)$ and show that $d_k(\cdot) \in \mathcal{D}$ for any $k \ge 2$, where $\mathcal{D}$ is as in Exercise 1.
3. Prove Vaughan's identity (9.4) and (9.5)–(9.8).

4. Using the notation introduced in the proof of Theorem 9.1.2, find a non-trivial upper estimate for

$$\sum_{d \leq z} \frac{d}{\phi(d)} \sideset{}{^*}\sum_{\chi} \max_{y \leq x} \left| \sum_{\substack{eh \leq y \\ U < h \leq UV}} b(e) \left(\sum_{\substack{f \leq V, g \leq U \\ fg = h}} \tilde{b}(f) c(g) \right) \chi(eh) \right|$$

by applying the modified large sieve inequality to the pair of sequences of complex numbers $(b(e))_{x/UV < e \leq x/U}$ and $(\tilde{b}(f)c(g))_{fg=h, U < h \leq UV}$. Compare the result with estimate (9.17).

5. Let $\zeta(s) = \sum_{n \geq 1} 1/n^s$ be the Riemann zeta function. Show that

$$-\frac{\zeta'(s)}{\zeta(s)} = \sum_{n \geq 1} \frac{\Lambda(n)}{n^s}$$

and

$$\frac{1}{\zeta(s)} = \sum_{n \geq 1} \frac{\mu(n)}{n^s}.$$

6. Prove Corollary 9.2.3.

7. Show that for any positive integer n,

$$d(n) = 2 \sum_{\substack{d \mid n \\ d \leq \sqrt{n}}} 1 - \delta(n),$$

where $\delta(n)$ is 1 if n is a square and 0 otherwise.

8. With the function $\delta(\cdot)$ defined as in the previous exercise, show that for any fixed integer a,

$$\sum_{p \leq x} \delta(p + a) = O\left(\sqrt{x}\right).$$

9. In the proof of Theorem 9.3.1 use the conditional asymptotic formula for $\pi(x; d, -a)$ (given in Section 7.3 of Chapter 7) instead of the Bombieri–Vinogradov theorem to prove (9.35).

10. Let $f(n)$ be an arithmetical function satisfying

$$\sum_{d > y} \frac{|f(d)|}{\phi(d)} \ll (\log y)^{-2}.$$

Let

$$g(n) := \sum_{d \mid n} f(d).$$

Show that, as x tends to infinity,

$$\sum_{p \leq x} g(p-1) = \left(\sum_{d=1}^{\infty} \frac{f(d)}{\phi(d)} \right) \operatorname{li} x + O(x \log^{-2} x),$$

where the summation is over primes p.

11. Show that the number of primes $p \leq x$ such that $p-1$ is squarefree is equal to

$$\left(\sum_{d=1}^{\infty} \frac{\mu(d)}{d\phi(d)} \right) \operatorname{li} x + O(x(\log^{-A} x))$$

for any $A > 1$. [Hint: observe that

$$\sum_{d^2 | n} \mu(d)$$

equals 1 if n is squarefree and zero otherwise.]

12. Prove that every natural number greater than 2 can be written as the sum of a prime and a squarefree number.

10

The lower bound sieve

In a sense, the first lower bound sieve was derived by Viggo Brun (1885–1978) in the year 1919. After he had introduced his Λ^2 method (Selberg's name for his sieve), Selberg indicated how his method can be developed into a lower bound sieve. The treatment we give here is due to Bombieri [1]. It will be seen that the sieve method is still of a combinatorial nature, suggesting its development and applicability to a wider context.

10.1 The lower bound sieve

In this section we exploit in more depth Selberg's idea discussed in Chapter 7 and developed as Selberg's sieve. The result that we will obtain is a more powerful sieve technique, called the **lower bound sieve**.

We consider our usual general setting: $\mathcal{A}$ is a finite set of positive integers; $\mathcal{P}$ is a set of rational primes; for each prime $p \in \mathcal{P}$, $\mathcal{A}_p$ is a subset of $\mathcal{A}$; $\mathcal{A}_1 := \mathcal{A}$ and for any squarefree positive integer d composed of primes of $\mathcal{P}$, $\mathcal{A}_d := \cap_{p|d} \mathcal{A}_p$; for a positive real number z, $P(z)$ denotes the product of the primes $p \in \mathcal{P}$ such that $p < z$.

In the previous chapters we were concerned with finding estimates for

$$S(\mathcal{A}, \mathcal{P}, z) := \# \left(\mathcal{A} \setminus \cup_{p|P(z)} \mathcal{A}_p \right).$$

We recall that the starting point in our combinatorial treatments of this problem was the formula

$$S(\mathcal{A}, \mathcal{P}, z) = \sum_{a \in \mathcal{A}} \sum_{\substack{d|P(z) \\ a \in \mathcal{A}_d}} \mu(d),$$

177

and that Selberg's original idea was the use in the above of the inequality

$$\sum_{\substack{d|P(z) \\ a \in \mathcal{A}_d}} \mu(d) \leq \left(\sum_{\substack{d|P(z) \\ a \in \mathcal{A}_d}} \lambda_d \right)^2,$$

valid for any sequence of real numbers (λ_d) satisfying $\lambda_1 = 1$.

We will elaborate on Selberg's idea and study the weighted sum

$$\sum_{a \in \mathcal{A}} \left(\sum_{\substack{t \\ a \in \mathcal{A}_t}} w_t \right) \left(\sum_{\substack{d \\ a \in \mathcal{A}_d}} \lambda_d \right)^2,$$

where (w_t) and (λ_d) are arbitrary sequences of real numbers supported only at positive squarefree integers t, d composed of primes of $\mathcal{P}$. Following E. Bombieri, we prove:

Theorem 10.1.1   *(The lower bound sieve)*
We keep the above setting and assume that there exist $X > 0$ and a multiplicative function $f(\cdot)$ such that, for any positive squarefree integer d composed of primes of $\mathcal{P}$,

$$\#\mathcal{A}_d = \frac{X}{f(d)} + R_d \qquad (10.1)$$

for some real number R_d. We write

$$f(n) = \sum_{d|n} f_1(d) \qquad (10.2)$$

for some multiplicative function $f_1(\cdot)$, uniquely determined by f. Then, for any $y, z > 0$ and for any sequences of real numbers $(w_t), (\lambda_d)$ that are supported only at positive squarefree integers $t \leq y, d \leq z$ composed of primes of $\mathcal{P}$, we have

$$\sum_{a \in \mathcal{A}} \left(\sum_{\substack{t \\ a \in \mathcal{A}_t}} w_t \right) \left(\sum_{\substack{d \\ a \in \mathcal{A}_d}} \lambda_d \right)^2 = \Delta X + E, \qquad (10.3)$$

where

$$E := O\left(\sum_{\substack{m \leq yz^2 \\ m|P(yz)}} \left(\sum_{\substack{t \leq y \\ t|m}} |w_t| \right) \left(\sum_{\substack{d \leq z \\ d|m}} |\lambda_d| \right)^2 |R_m| \right) \qquad (10.4)$$

and

$$\Delta := \sum_{\substack{t \le y, \delta \le z \\ r|P(y), \delta|P(z) \\ (t,\delta)=1}} \frac{w_t}{f(t)} \cdot \frac{1}{f_1(\delta)} \left(\sum_{\substack{r \le z/\delta \\ r|P(z) \\ r|t}} \mu(r) z_{\delta r} \right)^2, \qquad (10.5)$$

with

$$z_r := \mu(r) f_1(r) \sum_{\substack{s \le z/r \\ s|P(z)}} \frac{\lambda_{sr}}{f(sr)} \qquad (10.6)$$

for any positive squarefree integer r composed of primes of $\mathcal{P}$.

Remark 10.1.2 First we note that choosing the sequence (λ_d) is equivalent to choosing the sequence (z_r), since, by the dual Möbius inversion formula, (10.6) is equivalent to

$$\lambda_d = \mu(d) f(d) \sum_{\substack{r \le z/d \\ r|P(z)}} \frac{\mu^2(rd)}{f_1(rd)} z_{rd}. \qquad (10.7)$$

Then we note that Theorem 10.1.1 is a generalization of Selberg's sieve discussed in Chapter 7. To see this, we choose

$$w_t := \begin{cases} 1 & \text{if} \quad t = 1 \\ 0 & \text{if} \quad t > 1 \end{cases} \qquad (10.8)$$

and

$$z_r := \begin{cases} \frac{1}{V(z)} & \text{if} \quad r \le z \text{ and } r|P(z) \\ 0 & \text{otherwise}, \end{cases} \qquad (10.9)$$

where

$$V(z) := \sum_{\substack{d \le z \\ d|P(z)}} \frac{\mu^2(d)}{f_1(d)} \qquad (10.10)$$

(notice that this is the same as $V(z)$ defined in Chapter 7). On one hand, from (10.7) and (10.9) we get that

$$\lambda_d = \frac{1}{V(z)} \cdot \frac{\mu(d) f(d)}{f_1(d)} \sum_{\substack{r \le z/d \\ r|P(z) \\ (r,d)=1}} \frac{\mu^2(r)}{f_1(r)}, \qquad (10.11)$$

which is exactly what we had chosen in the proof of Selberg's sieve. For this choice, we have $\lambda_d = 0$ if $d > z$, $\lambda_1 = 1$ and $|\lambda_d| \leq 1$ for any $d \leq z$. Using these observations in (10.4) we obtain that

$$E = O\left(\sum_{\substack{d_1, d_2 \leq z \\ d_1, d_2 | P(z)}} |R_{[d_1, d_2]}| \right).$$

On the other hand, from (10.9) and (10.5) we get

$$\Delta = \frac{1}{V(z)}.$$

Thus

$$S(\mathcal{A}, \mathcal{P}, z) \leq \frac{X}{V(z)} + O\left(\sum_{\substack{d_1, d_2 \leq z \\ d_1, d_2 | P(z)}} |R_{[d_1, d_2]}| \right),$$

which is exactly Selberg's sieve.

The proof of Theorem 10.1.1 is a 'weighted replica' of the treatment of the summation

$$\sum_{a \in \mathcal{A}} \left(\sum_{\substack{d \\ a \in \mathcal{A}_d}} \lambda_d \right)^2$$

occuring in Selberg's sieve. We will need the following auxiliary results.

Lemma 10.1.3 *Let f be a multiplicative function. For any squarefree positive integers t, d_1, d_2 we have*

$$f[t, d_1, d_2] = \frac{f(t) f(d_1) f(d_2) f(t, d_1, d_2)}{f(t, d_1) f(t, d_2) f(d_1, d_2)},$$

where, for integers a and b, $f[a, b] = f([a, b])$ and $f(a, b) = f((a, b))$.

Proof Exercise. $\square$

Lemma 10.1.4 *Let f be a multiplicative function, let f_1 be defined by (10.2) and let f_{-1} be defined by*

$$\frac{1}{f(n)} = \sum_{d | n} f_{-1}(d). \qquad (10.12)$$

Let t, δ be squarefree positive integers. Then

$$\sum_{\substack{[r,s]=\delta \\ s|t}} f_1(r)f_{-1}(s) = \begin{cases} f_1(\delta) & \text{if } (t,\delta)=1 \\ 0 & \text{otherwise.} \end{cases}$$

Proof Let r, s be squarefree positive integers, and let $\tau := [r,s]$, $\delta := (r,s)$. Using Lemma 10.1.3 we see that

$$f_1(\delta) = \frac{f_1(r)f_1(\tau)}{f_1(\tau)},$$

or, equivalently,

$$f_1(r) = \frac{f_1(\delta)f_1(\tau)}{f_1(s)}.$$

This allows us to write

$$\sum_{\substack{[r,s]=\delta \\ s|t}} f_1(r)f_{-1}(s) = f_1(\delta) \sum_{s|(t,\delta)} \frac{f_{-1}(s)}{f_1(s)} \sum_{\tau|s} f_1(\tau)$$

$$= f_1(\delta) \sum_{s|(t,\delta)} \frac{f_{-1}(s)f(s)}{f_1(s)}.$$

Now we notice that for any prime p we have

$$\frac{f_{-1}(p)f(p)}{f_1(p)} = \frac{\left(\frac{1}{f(p)}-1\right)f(p)}{f(p)-1} = -1.$$

Since $f_{-1}(\cdot)f(\cdot)/f_1(\cdot)$ is a multiplicative function, we must have

$$\frac{f_{-1}(s)f(s)}{f_1(s)} = \mu(s).$$

Therefore

$$\sum_{\substack{[r,s]=\delta \\ s|t}} f_1(r)f_{-1}(s) = f_1(\delta) \sum_{s|(t,\delta)} \mu(s),$$

which, by Lemma 1.2.1 of Chapter 1, gives us the result claimed in the lemma. $\square$

Lemma 10.1.5 *Let f be a multiplicative function and let f_1 be defined by (10.2). Let r, t, δ be positive integers such that r is squarefree and $(t, \delta) = 1$. Then*

$$\sum_{s|r} \mu\left(\frac{r}{s}\right) f(t, \delta s) = \begin{cases} f_1(r) & \text{if } r|t \\ 0 & \text{if } r \nmid t. \end{cases}$$

Proof   First we observe that since $(t, \delta) = 1$, we have $(t, \delta s) = (t, s)$. Thus we need to analyze

$$S := \sum_{s|r} \mu\left(\frac{r}{s}\right) f(t, s).$$

Let $r = p_1 \ldots p_k$ be the prime factorization of r. First let us consider the case when $r|t$, hence the case when $t = p_1 \ldots p_k t'$ for some positive integer t'. Then

$$S = \sum_{\substack{0 \le \alpha_1 \le 1 \\ \cdots \\ 0 \le \alpha_k \le 1}} \mu\left(p_1^{1-\alpha_1} \ldots p_k^{1-\alpha_k}\right) f\left(p_1^{\alpha_1} \ldots p_k^{\alpha_k}\right)$$

$$= \prod_{1 \le j \le k} (f(p_j) - 1)$$

$$= f_1(r).$$

Now let us consider the case when $r \nmid t$. Without loss of generality we can write $t = p_i \ldots p_k t'$ for some $i > 1$ and some integer t' coprime to $p_1 \ldots p_{i-1}$. Then

$$S = \sum_{\substack{0 \le \alpha_i \le 1 \\ \cdots \\ 0 \le \alpha_k \le 1}} \mu\left(p_1 \ldots p_{i-1} p_i^{1-\alpha_i} \ldots p_k^{1-\alpha_k}\right) f\left(p_i \ldots p_k t', p_1^{\alpha_1} \ldots p_k^{\alpha_k}\right)$$

$$= \prod_{1 \le j \le i-1} \sum_{0 \le \alpha_j \le 1} \mu\left(p_j^{1-\alpha_j}\right) \cdot \prod_{i \le j \le k} \sum_{0 \le \alpha_j \le 1} \mu\left(p_j^{1-\alpha_j}\right) f\left(p_j^{\alpha_j}\right)$$

$$= \prod_{1 \le j \le i-1} (\mu(p_j) + 1) \cdot \prod_{i \le j \le k} \sum_{0 \le \alpha_j \le 1} \mu\left(p_j^{1-\alpha_j}\right) f\left(p_j^{\alpha_j}\right)$$

$$= 0.$$

This completes the proof of the lemma. $\square$

Proof of Theorem 10.1.1   By squaring out the sum over d, interchanging summation and using the assumption on $\#\mathcal{A}_d$, we obtain

$$\sum_{a \in \mathcal{A}} \left(\sum_{\substack{t \le y \\ t|P(y) \\ a \in \mathcal{A}_t}} w_t \right) \left(\sum_{\substack{d \le z \\ d|P(z) \\ a \in \mathcal{A}_d}} \lambda_d \right)^2$$

$$= X \sum_{\substack{t \le y; d_1, d_2 \le z \\ t|P(y); d_1, d_2|P(z)}} \frac{w_t \lambda_{d_1} \lambda_{d_2}}{f[t, d_1, d_2]} + \sum_{\substack{t \le y; d_1, d_2 \le z \\ t|P(y); d_1, d_2|P(z)}} w_t \lambda_{d_1} \lambda_{d_2} R_{[t, d_1, d_2]}$$

$$= X \sum_{\substack{t \le y; d_1, d_2 \le z \\ t|P(y); d_1, d_2|P(z)}} \frac{w_t \lambda_{d_1} \lambda_{d_2}}{f[t, d_1, d_2]}$$

$$+ O\left(\sum_{\substack{m \le yz^2 \\ m|P(yz)}} \left(\sum_{\substack{t \le y \\ t|m}} |w_t| \right) \left(\sum_{\substack{d \le z \\ d|m}} |\lambda_d| \right)^2 |R_m| \right). \tag{10.13}$$

Now we analyze

$$\Delta' := \sum_{\substack{t \le y; d_1, d_2 \le z \\ t|P(y); d_1, d_2|P(z)}} \frac{w_t \lambda_{d_1} \lambda_{d_2}}{f[t, d_1, d_2]} \tag{10.14}$$

which, by using Lemma 10.1.3 and formulae (10.2) and (10.12), can be rewritten as

$$= \sum_{\substack{t \le y; d_1, d_2 \le z \\ t|P(y); d_1, d_2|P(z)}} \frac{w_t}{f(t)} \cdot \frac{\lambda_{d_1} f(t, d_1)}{f(d_1)} \cdot \frac{\lambda_{d_2} f(t, d_2)}{f(d_2)} \left(\sum_{r|(d_1, d_2)} f_1(r) \right) \left(\sum_{s|(t, d_1, d_2)} f_{-1}(s) \right)$$

$$= \sum_{\substack{t \le y; r, s \le z \\ t|P(y); r, s|P(z) \\ s|t}} \frac{w_t}{f(t)} f_1(r) f_{-1}(s) \sum_{\substack{d_1, d_2 \le z \\ d_1, d_2|P(z) \\ [r,s]|d_1 \\ [r,s]|d_2}} \frac{\lambda_{d_1} f(t, d_1)}{f(d_1)} \cdot \frac{\lambda_{d_2} f(t, d_2)}{f(d_2)}$$

$$= \sum_{\substack{t \le y; r, s \le z \\ t|P(y); r, s|P(z) \\ s|t}} \frac{w_t}{f(t)} f_1(r) f_{-1}(s) \left(\sum_{\substack{d \le z \\ d|P(z) \\ [r,s]|d}} \frac{\lambda_d f(t, d)}{f(d)} \right)^2.$$

For any t and δ with $t|P(y)$ and $\delta|P(z)$ let

$$u_{t, \delta} := \sum_{\substack{d \le z \\ d|P(z) \\ \delta|d}} \frac{\lambda_d f(t, d)}{f(d)}. \tag{10.15}$$

By the dual Möbius inversion formula this is equivalent to

$$\frac{\lambda_\delta f(t, \delta)}{f(\delta)} = \sum_{\substack{d \le z \\ d|P(z) \\ \delta|d}} \mu\left(\frac{d}{\delta}\right) u_{t, d}. \tag{10.16}$$

By using (10.15) and Lemma 10.1.4 we obtain

$$\Delta' = \sum_{\substack{t \le y; \delta \le z \\ t|P(y); \delta|P(z)}} \frac{w_t}{f(t)} u_{t,\delta}^2 \sum_{\substack{[r,s]=\delta \\ s|t}} f_1(r) f_{-1}(s)$$

$$= \sum_{\substack{t \le y; \delta \le z \\ t|P(y); \delta|P(z) \\ (t,\delta)=1}} \frac{w_t}{f(t)} f_1(\delta) u_{t,\delta}^2. \tag{10.17}$$

We note that (10.17) is reminiscent of (7.18) of Chapter 7, since in the situation of Selberg's sieve the sequence (w_t) is given by $w_1 = 1$ and $w_t = 0$ for any $t > 1$ (see Remark 10.1.2). If we view Δ' as a quadratic form in (λ_d), then we see that, through the above calculations, we have diagonalized this form.

In what follows we will try to bring the summation

$$\sum_{\substack{\delta \le z \\ \delta|P(z)}} \frac{1}{f_1(\delta)}$$

into the expression of Δ'. This will emphasize the resemblance of the result discussed here with Selberg's sieve. Let us note that from (10.15) and (10.16) we get

$$u_{1,\delta} = \sum_{\substack{m \le z/\delta \\ m|P(z)}} \frac{\lambda_{\delta m}}{f(\delta m)} \tag{10.18}$$

and

$$\frac{\lambda_\delta}{f(\delta)} = \sum_{\substack{t \le z/\delta \\ t|P(z)}} \mu(t) u_{1,\delta t}. \tag{10.19}$$

By using (10.19) and Lemma 10.1.5 we rewrite $u_{t,\delta}$ as

$$u_{t,\delta} = \sum_{\substack{s \le z/\delta \\ s|P(z)}} \frac{\lambda_{\delta s} f(t, \delta s)}{f(\delta s)} = \sum_{\substack{s \le z/\delta \\ s|P(z)}} f(t, \delta s) \sum_{\substack{m \le z/(\delta s) \\ m|P(z)}} \mu(m) u_{1,\delta s m}$$

$$= \sum_{\substack{r \le z/\delta \\ r|P(z)}} \left(\sum_{s|r} \mu\left(\frac{r}{s}\right) f(t, \delta s) \right) u_{1,\delta r} = \sum_{\substack{r \le z/\delta \\ r|P(z) \\ r|t}} f_1(r) u_{1,\delta r}.$$

Now we use (10.18) and the observation that if $r|t$ and $(t, \delta) = 1$, then $(r, \delta) = 1$, to write

$$u_{t,\delta} = \sum_{\substack{r \leq z/\delta \\ r|P(z) \\ r|t}} f_1(r) \sum_{\substack{s \leq z/(\delta r) \\ s|P(z)}} \frac{\lambda_{\delta rs}}{f(\delta rs)}$$

$$= \frac{\mu(\delta)}{f_1(\delta)} \sum_{\substack{r \leq z/\delta \\ r|P(z) \\ r|t}} \mu(r)\mu(\delta r)f_1(\delta r) \sum_{\substack{s \leq z/(\delta r) \\ s|P(z)}} \frac{\lambda_{\delta rs}}{f(\delta rs)}.$$

We plug this back into (10.17) and get

$$\Delta' = \sum_{\substack{t \leq y; \delta \leq z \\ t|P(y); \delta|P(z) \\ (t,\delta)=1}} \frac{w_t}{f(t)} \cdot \frac{1}{f_1(\delta)} \left(\sum_{\substack{r \leq z/\delta \\ r|P(z) \\ r|t}} \mu(r)\mu(\delta r)f_1(\delta r) \sum_{\substack{s \leq z/(\delta r) \\ s|P(z)}} \frac{\lambda_{\delta rs}}{f(\delta rs)} \right)^2 .$$

By introducing the notation

$$z_r := \mu(r)f_1(r) \sum_{\substack{s \leq z/r \\ s|P(z)}} \frac{\lambda_{sr}}{f(sr)},$$

we finally obtain

$$\Delta' = \sum_{\substack{t \leq y; \delta \leq z \\ t|P(y); \delta|P(z) \\ (t,\delta)=1}} \frac{w_t}{f(t)} \cdot \frac{1}{f_1(\delta)} \left(\sum_{\substack{r \leq z/\delta \\ r|P(z) \\ r|t}} \mu(r)z_{\delta r} \right)^2 ,$$

which is exactly the quantity Δ introduced in the statement of the theorem. This completes the proof. $\square$

10.2 Twin primes

In 1849, de Polignac conjectured that every even number is the difference of two primes in infinitely many ways. In particular, the conjecture predicts that there are infinitely many primes p such that $p+2$ is also prime. This special case of de Polignac's conjecture undoubtedly has roots in antiquity. In this section, we will prove:

Theorem 10.2.1   *There are infinitely many primes p such that p+2 has at most four distinct prime factors.*

This is an application of the lower bound sieve discussed in Section 10.1 and of the following preliminary lemmas.

Lemma 10.2.2   *(Mertens' formula)*

$$\sum_{p \le x} \frac{1}{p-1} = \log(1 + \log x) + A_0 + O\left(\frac{1}{1+\log x}\right)$$

for some positive constant A_0.

Proof   Exercise. □

We denote by $\phi_1(\cdot)$ the multiplicative function defined by $\phi(n) = \sum_{d|n} \phi_1(d)$.

Lemma 10.2.3

$$\sum_{\substack{\delta \le x \\ 2 \nmid \delta}} \frac{\mu^2(\delta)}{\phi_1(\delta)} = A_1 \log x + A_2 + O\left(\frac{1}{x^{1/4}}\right)$$

for some positive constants A_1, A_2.

Proof   Exercise. □

Lemma 10.2.4

$$\sum_{\substack{\delta \le x \\ 2 \nmid \delta}} \frac{\mu^2(\delta)}{\phi_1(\delta)} \cdot \frac{1}{1+\log \frac{x}{\delta}} \ll \log \log x.$$

Proof   We break the interval $(2, x]$ into dyadic subintervals of the form $\left(x/2^{k+1}, x/2^k\right)$ with $0 \le k < (\log x/\log 2) - 2$. Then

$$\sum_{\substack{\delta \le x \\ 2 \nmid \delta}} \frac{\mu^2(\delta)}{\phi_1(\delta)} \cdot \frac{1}{1+\log \frac{x}{\delta}} \ll \sum_{1 \le k < \log x} \frac{1}{k} \sum_{\frac{x}{2^{k+1}} < \delta \le \frac{x}{2^k}} \frac{\mu^2(\delta)}{\phi_1(\delta)}.$$

By using Lemma 10.2.3, the above becomes

$$\ll \sum_{1 \le k < \log x} \frac{1}{k}\left(1 + \left(\frac{2^k}{x}\right)^{1/4}\right) \ll \sum_{1 \le k < \log x} \frac{1}{k} \ll \log \log x.$$

This completes the proof of the lemma. □

Lemma 10.2.5

$$\sum_{\substack{\delta \le x \\ 2 \nmid \delta}} \frac{\mu^2(\delta)}{\phi_1(\delta)} \sum_{\substack{\frac{x}{\delta} < q \le y \\ q \mid 2\delta}} \frac{1}{q-1} = O(1).$$

Proof We write

$$\sum_{\substack{\delta \le x \\ 2 \nmid \delta}} \frac{\mu^2(\delta)}{\phi_1(\delta)} \sum_{\substack{\frac{x}{\delta} < q \le y \\ q \mid 2\delta}} \frac{1}{q-1} = \sum_{q \le y} \frac{1}{q-1} \sum_{\substack{\frac{x}{q} < \delta \le x \\ 2 \nmid \delta \\ q \mid 2\delta}} \frac{\mu^2(\delta)}{\phi_1(\delta)}$$

$$= \sum_{\substack{\frac{x}{2} < \delta \le x \\ 2 \nmid \delta}} \frac{\mu^2(\delta)}{\phi_1(\delta)} + \sum_{2 \ne q \le y} \frac{1}{q-1} \sum_{\substack{\frac{x}{q} < \delta \le x \\ q \mid \delta}} \frac{\mu^2(\delta)}{\phi_1(\delta)}$$

$$= \sum_{\substack{\frac{x}{2} < \delta \le x \\ 2 \nmid \delta}} \frac{\mu^2(\delta)}{\phi_1(\delta)} + \sum_{2 \ne q \le y} \frac{1}{(q-1)(q-2)} \sum_{\substack{\frac{x}{q^2} < \delta_1 \le \frac{x}{q} \\ q \nmid \delta_1}} \frac{\mu^2(\delta_1)}{\phi_1(\delta_1)}.$$

Now we apply Lemma 10.2.3 and obtain

$$\sum_{\substack{\delta \le x \\ 2 \nmid \delta}} \frac{\mu^2(\delta)}{\phi_1(\delta)} \sum_{\substack{\frac{x}{\delta} < q \le y \\ q \mid 2\delta}} \frac{1}{q-1} = O\left(\frac{1}{x^{1/4}}\right) + O(1) = O(1).$$

$\square$

Lemma 10.2.6 *As $x \to \infty$,*

$$\sum_{\substack{\delta \le x \\ 2 \nmid \delta}} \frac{\mu^2(\delta)}{\phi_1(\delta)} \log \frac{\log x}{1 + \log \frac{x}{\delta}} \sim \sum_{\substack{\delta \le x \\ 2 \nmid \delta}} \frac{\mu^2(\delta)}{\phi_1(\delta)}.$$

Proof We apply partial summation and then Lemma 10.2.3 to obtain

$$\sum_{\substack{\delta \le x \\ 2 \nmid \delta}} \frac{\mu^2(\delta)}{\phi_1(\delta)} \log \frac{\log x}{1 + \log \frac{x}{\delta}} = (\log \log x) \sum_{\substack{\delta \le x \\ 2 \nmid \delta}} \frac{\mu^2(\delta)}{\phi_1(\delta)} - \sum_{\substack{\delta \le x \\ 2 \nmid \delta}} \frac{\mu^2(\delta)}{\phi_1(\delta)} \log \left(1 + \log \frac{x}{\delta}\right)$$

$$= A_1(\log x)(\log \log x) + A_2 \log \log x$$

$$+ O\left(\frac{\log \log x}{x^{1/4}}\right) - I,$$

where

$$I := \int_3^x \left(\sum_{\substack{\delta \leq t \\ 2 \nmid \delta}} \frac{\mu^2(\delta)}{\phi_1(\delta)} \right) \frac{1}{t\left(1 + \log \frac{x}{t}\right)} \, dt$$

and where the constants A_1, A_2 are as in Lemma 10.2.3. To handle the integral I we make use once more of Lemma 10.2.3. We obtain

$$I = -A_1 \int_3^x \frac{dt}{t} + (A_1 \log x + A_1 + A_2) \int_3^x \frac{dt}{t(1 + \log x - \log t)} + O\left(\frac{1}{x^{1/4}}\right)$$

$$= -A_1 \log x + A_1 \log 3 + A_1 (\log x)(\log \log x)$$

$$+ (A_1 + A_2) \log \log x + O\left(\frac{1}{\log x}\right).$$

Putting everything together gives us that

$$\sum_{\substack{\delta \leq x \\ 2 \nmid \delta}} \frac{\mu^2(\delta)}{\phi_1(\delta)} \log \frac{\log x}{1 + \log \frac{x}{\delta}} \sim A_1 \log x,$$

which, by Lemma 10.2.3, completes the proof of the lemma. $\square$

Lemma 10.2.7   *Let k be a positive integer. Then*

$$\sum_{m \leq x} \frac{d^k(m)}{m} \ll (\log x)^{2^k}.$$

Proof   Exercise. $\square$

Proof of Theorem 10.2.1   Let

$$\mathcal{A} := \{p + 2 : p \leq x, p \text{ prime}\},$$

$$\mathcal{P} := \{q : q \neq 2, q \text{ prime}\},$$

and for each $q \in \mathcal{P}$, let

$$\mathcal{A}_q := \{p + 2 : p \leq x, p \text{ prime}, q | p + 2\}.$$

We observe that for any squarefree positive integer d composed of primes of $\mathcal{P}$ we have

$$\#\mathcal{A}_d = \pi(x; d, -2) = \frac{1}{\phi(d)} \mathrm{li}\, x + R_d \qquad (10.20)$$

for some R_d satisfying $dR_d \ll x$.

By applying Theorem 10.1.1 we get that for any $y, z > 0$ and for any sequences of real numbers (w_t) and (λ_d) we have

$$\sum_{p \leq x} \left(\sum_{\substack{t \leq y \\ t \mid P(y) \\ t \mid (p+2)}} w_t \right) \left(\sum_{\substack{d \leq z \\ d \mid P(z) \\ d \mid (p+2)}} \lambda_d \right)^2 = \Delta \operatorname{li} x + E \qquad (10.21)$$

where Δ and E are as in (10.5) and (10.4), respectively, with the multiplicative function $f(\cdot)$ replaced by the Euler function $\phi(\cdot)$.

The parameters y and z depend on and grow with x, and will be specified later.

We choose the sequence (λ_d) as in Selberg's sieve, that is, we choose (z_r) according to (10.9) and, equivalently, (λ_d) according to (10.11). This implies that

$$E = O\left(\sum_{\substack{m \leq yz^2 \\ m \mid P(yz)}} \left(\sum_{\substack{t \leq y \\ t \mid m}} |w_t| \right) d^2(m) |R_m| \right), \qquad (10.22)$$

where $d(\cdot)$ is the divisor function, and that

$$\Delta = \frac{1}{V(z)^2} \sum_{\substack{t \leq y, \delta \leq z \\ t \mid P(y), \delta \mid P(z) \\ (t,\delta)=1}} \frac{w_t}{\phi(t)} \cdot \frac{1}{\phi_1(\delta)} \left(\sum_{\substack{r \leq z/\delta \\ r \mid P(z) \\ r \mid t}} \mu(r) \right)^2, \qquad (10.23)$$

where

$$V(z) := \sum_{\substack{d \leq z \\ d \mid P(z)}} \frac{1}{\phi_1(d)} = \sum_{\substack{d \leq z \\ 2 \nmid d}} \frac{\mu^2(d)}{\phi_1(d)}. \qquad (10.24)$$

The sequence (w_t) will be chosen in two different ways, as follows.

The first choice of (w_t) We set

$$w_t := \begin{cases} 1 & \text{if } t = 1 \\ 0 & \text{if } t > 1. \end{cases} \qquad (10.25)$$

Then (10.22) becomes

$$E_1 = O\left(\sum_{m \leq z^2} d^2(m) |R_m| \right) \qquad (10.26)$$

and (10.23) becomes

$$\Delta_1 = \frac{1}{V(z)^2} \sum_{\substack{\delta \leq z \\ \delta | P(z)}} \frac{1}{\phi_1(\delta)} = \frac{1}{V(z)}. \tag{10.27}$$

In other words,

$$\sum_{p \leq x} \left(\sum_{\substack{d \leq z \\ 2 \nmid d; \mu^2(d)=1 \\ d|(p+2)}} \lambda_d \right)^2 = \frac{1}{V(z)} \operatorname{li} x + O\left(\sum_{m \leq z^2} d^2(m)|R_m| \right). \tag{10.28}$$

The second choice of (w_t) We set

$$w_t := \begin{cases} 1 & \text{if } t \leq y \text{ and } t \text{ prime} \\ 0 & \text{otherwise.} \end{cases} \tag{10.29}$$

Then (10.22) becomes

$$E_2 = O\left(\sum_{m \leq yz^2} d^3(m)|R_m| \right) \tag{10.30}$$

and, by using Mertens' formula, (10.23) becomes

$$\Delta_2 = \frac{1}{V(z)^2} \sum_{\substack{q \leq y, \delta \leq z \\ \delta | P(z) \\ \delta q > z, q \nmid 2\delta}} \frac{1}{(q-1)\phi_1(\delta)}$$

$$= \frac{1}{V(z)^2} \sum_{\substack{\delta \leq z \\ 2 \nmid \delta}} \frac{\mu^2(\delta)}{\phi_1(\delta)} \left(\sum_{\frac{z}{\delta} < q \leq y} \frac{1}{q-1} - \sum_{\substack{\frac{z}{\delta} < q \leq y \\ q | 2\delta}} \frac{1}{q-1} \right)$$

$$= \frac{1}{V(z)^2} \sum_{\substack{\delta \leq z \\ 2 \nmid \delta}} \frac{\mu^2(\delta)}{\phi_1(\delta)} \left(\log \frac{1+\log y}{1+\log \frac{z}{\delta}} + O\left(\frac{1}{1+\log \frac{z}{\delta}} \right) - \sum_{\substack{\frac{z}{\delta} < q \leq y \\ q | 2\delta}} \frac{1}{q-1} \right)$$

$$= \frac{1}{V(z)} \log \frac{1+\log y}{\log z}$$

$$+ \frac{1}{V(z)^2} \sum_{\substack{\delta \leq z \\ 2 \nmid \delta}} \frac{\mu^2(\delta)}{\phi_1(\delta)} \left(\log \frac{\log z}{1+\log \frac{z}{\delta}} + O\left(\frac{1}{1+\log \frac{z}{\delta}} \right) - \sum_{\substack{\frac{z}{\delta} < q \leq y \\ q | 2\delta}} \frac{1}{q-1} \right).$$

By using Lemmas 10.2.4–10.2.6 we deduce that

$$\Delta_2 = \frac{1+\log\left(\frac{\log y}{\log z}\right)}{V(z)} + o\left(\frac{1}{V(z)}\right). \tag{10.31}$$

Therefore

$$\sum_{p\leq x}\left(\sum_{\substack{2\neq q\leq y \\ q\mid(p+2)}}1\right)\left(\sum_{\substack{d\leq z \\ 2\nmid d;\mu^2(d)=1 \\ d\mid(p+2)}}\lambda_d\right)^2 = \frac{1+\log\frac{\log y}{\log z}}{V(z)}\operatorname{li}x + o\left(\frac{x}{V(z)\log x}\right)$$

$$+ O\left(\sum_{m\leq yz^2}d^3(m)|R_m|\right). \tag{10.32}$$

Now we want to choose y and z such that $E_1 = E_2 = o\left(x/(\log x)^2\right)$. By using the Cauchy–Schwarz inequality and Lemma 10.2.7 in (10.26) and (10.30) we obtain

$$E_1 \ll \left(\sum_{m\leq z^2}\frac{d^4(m)}{m}\right)^{1/2}\left(\sum_{m\leq z^2}m|R_m|^2\right)^{1/2} \ll (\log z)^\alpha x^{1/2}\left(\sum_{m\leq z^2}|R_m|\right)^{1/2}$$

for some $\alpha > 0$, where we have also used that $m|R_m| \ll x$ for any m. Similarly,

$$E_2 \ll (\log yz)^\beta x^{1/2}\left(\sum_{m\leq yz^2}|R_m|\right)^{1/2}$$

for some $\beta > 0$. In order to estimate the summations $\sum_m|R_m|$, we invoke the Bombieri–Vinogradov theorem. To do this, we need

$$yz^2 \leq \frac{x^{1/2}}{(\log x)^B}$$

for a sufficiently large positive integer B. This condition is satisfied if we choose

$$y := x^{1/4}(\log x)^B \text{ and } z := \frac{x^{1/8}}{(\log x)^B}. \tag{10.33}$$

We then obtain

$$E_1 = E_2 = o\left(\frac{x}{(\log x)^2}\right). \tag{10.34}$$

Finally we combine (10.28), (10.32), (10.33) and (10.34) to deduce that, as $x \to \infty$,

$$\sum_{p \le x} \left(\sum_{\substack{d \le x^{1/8}(\log x)^{-B} \\ 2\nmid d;\, \mu^2(d)=1 \\ d|(p+2)}} \lambda_d \right)^2 \left(2 - \sum_{\substack{2 \ne q \le x^{1/4}(\log x)^B \\ q|(p+2)}} 1 \right) \sim \frac{1 - \log\left(\frac{\frac{1}{4}\log x + B \log \log x}{\frac{1}{8}\log x - B \log \log x} \right)}{V(x^{1/8}(\log x)^{-B})} \operatorname{li} x.$$

We notice that, as $x \to \infty$, the quantity on the right-hand side of the above asymptotic formula is positive, and so there must be infinitely many primes $p \le x$ such that

$$2 - \sum_{\substack{2 \ne q \le x^{1/4}(\log x)^B \\ q|(p+2)}} 1 > 0.$$

In other words, as $x \to \infty$, there are infinitely many primes $p \le x$ such that $p+2$ has at most one (odd) prime factor $\le x^{1/4}(\log x)^B$. Clearly, if $p \le x$, then $p+2$ has at most three prime factors $> x^{1/4}(\log x)^B$. Thus we proved the existence of infinitely many primes p for which $p+2$ has at most four prime divisors. □

We will now give a brief outline of a recent breakthrough in the theory of the lower bound sieve method. Let $\varpi(n)$ be $\log n$ if n is prime and zero otherwise. Let $H = \{h_1, \ldots, h_k\}$ be a set of k distinct integers. We now define the polynomial

$$P_H(n) = (n+h_1)(n+h_2)\cdots(n+h_k).$$

Fix a non-negative integer ℓ and let $R > 0$. In a recent paper soon to be published, D. Goldston, J. Pintz and C. Yildirim consider the sum

$$S(N) = \sum_{N \le n \le 2N} \left(\sum_{h \in H} \varpi(n+h) - \log 3N \right) \left(\sum_{d|P_H(n)} \lambda_d \right)^2$$

for some suitable choice of λ_d's. Note that, if $S(N)$ is positive for N sufficiently large, then there exist distinct h_i and h_j such that $n+h_i$ and $n+h_j$ are both prime for some n in the interval $[N, 2N]$. Goldston *et al.* choose λ_d to be

$$\mu(d) \frac{(\log R/d)^{k+\ell}}{(k+\ell)!}$$

for $d < R$ and zero otherwise. Then, for suitable choices of ℓ and R they show that $S(N)$ is strictly positive for N sufficiently large. Assuming the Halberstam–Elliott conjecture with exponent θ (see Section 9.2), they show that for $\theta > 1/2$ there is a constant $c(\theta)$ so that

$$\liminf_{n \to \infty} (p_{n+1} - p_n) \le c(\theta).$$

[Here p_n denotes the n-th prime.] If one can choose $\theta > 20/21$, then

$$\liminf_{n \to \infty} (p_{n+1} - p_n) \le 20.$$

Unconditionally, using the Bombieri–Vinogradov theorem, they show that

$$\liminf_{n \to \infty} \frac{(p_{n+1} - p_n)}{\log p_n} = 0,$$

settling a long-standing conjecture. This work represents a new approach to the twin prime problem. Undoubtedly, the methods will have other applications.

10.3 Quantitative results and variations

We begin by making some remarks on the calculation of the previous section. The first question we address is whether we can obtain a quantitative estimate. As the λ_ds satisfy $|\lambda_d| \le 1$, we immediately see that

$$\sum_{p \le x}' d(p+2)^2 \gg \frac{x}{\log^2 x}$$

where the dash on the summation indicates that

$$\sum_{\substack{q \le x^{1/4} (\log x)^B \\ q | p+2}} 1 < 2.$$

Thus, for each prime in the dashed summation, the number of prime divisors of $p+2$ is bounded. Consequently, $d(p+2)$ is bounded for these primes. We immediately deduce:

Theorem 10.3.1 *Let B be as in the previous section. The number of primes $p \le x$ having at most one prime factor $\le x^{1/4} (\log x)^B$ is*

$$\gg \frac{x}{(\log x)^2}.$$

It should now be clear that the method can be applied to treat other problems of a similar nature. For example, one can discuss a variation of the Goldbach conjecture and show that every sufficiently large even number can be written as a sum of a prime and another number which has at most four prime divisors. We leave the details to the reader.

One particular application we will highlight here. This concerns the problems of finding many primes p such that $(p-1)/2$ has few prime factors, all of them being very large. It is evident that the method of the previous section can be applied. We will only indicate the necessary modifications here and leave the details to the reader as an instructive exercise.

First, we sieve the sequence

$$\mathcal{A} := \{ (p-1)/2 : p \le x, p \text{ prime} \},$$

with

$$\mathcal{P} := \{ q : q \ne 2, q \text{ prime} \}.$$

The method of the previous section can be applied mutatis mutandis to deduce that there are

$$\gg \frac{x}{(\log x)^2}$$

primes $p \le x$ such that $(p-1)/2$ has at most four prime factors. However, with a view to applications in the next section, we will choose our parameters slightly differently.

The first condition to ensure is

$$yz^2 \le x^{1/2}/(\log x)^B,$$

with notation as in Section 10.2. Then, (10.34) is guaranteed. This allows us to deduce that

$$\sum_{p \le x} \left(\sum_{\substack{d \le z \\ d|(p-1)/2}} \lambda_d \right)^2 \left(1 - \sum_{\substack{q \le y \\ q|(p-1)/2}} 1 \right) \sim \frac{\log\left(\frac{\log z}{\log y}\right)}{V(z)} \operatorname{li} x.$$

We choose $z := x^{1/6}(\log x)^B$, $y := x^{1/6}/(\log x)^{2B}$. This enables us to obtain:

Theorem 10.3.2 *Let B be chosen as before. Then the number of primes $p \le x$ such that all the prime factors of $(p-1)/2$ are greater than $x^{1/6}/(\log x)^{2B}$ is*

$$\gg \frac{x \log\log x}{(\log x)^3}.$$

It is clear that the theorem affords other variations. We could, for example, have all the primes in a fixed residue class modulo some specifed modulus and still obtain an analogous result. It is also clear that we may suppose that all the prime divisors of $(p-1)/2$ are distinct. Indeed, the number of primes $p \le x$ for which $(p-1)/2$ is divisible by q^2 with $q > x^{1/6}/(\log x)^{2B}$ cannot exceed

$$\sum_{q > x^{1/6}/(\log x)^{2B}} \frac{x}{q^2} \ll x^{5/6}(\log x)^{2B},$$

which is negligible compared to the lower estimate provided by Theorem 10.3.2.

10.4 Application to primitive roots

In 1927, Emil Artin conjectured that any non-zero integer a other than ± 1 or a perfect square represents a generator of the coprime residue classes mod p for infinitely many primes p. This is known as **Artin's conjecture on primitive roots** (see [44] for history and details).

In 1967, C. Hooley [31] applied the sieve of Eratosthenes in conjunction with the generalized Riemann hypothesis to show that the number of primes $p \leq x$ for which a is a primitive root modulo p is

$$\sim A(a)\frac{x}{\log x}$$

as x tends to infinity and with $A(a) > 0$ whenever $a \neq \pm 1$ or a perfect square. It is still unknown if this can be obtained without the generalized Riemann hypothesis.

In 1984, Rajiv Gupta and Ram Murty [23] proved unconditionally that there is a finite set S such that for some $a \in S$, Artin's conjecture is true. Their work employed the lower bound sieve (in the form given by Iwaniec [34]) in conjunction with the Bombieri–Vinogradov theorem. As described in Chapter 9, the Bombieri–Vinogradov theorem has since been improved in various ways. Thus, if we inject the work of [2] into the arguments, one obtains a corresponding improvement in the results of [23]. This was done by Heath–Brown [25] and the best result to date is:

Theorem 10.4.1 *Let a_1, a_2, a_3 be three non-zero integers unequal to ± 1 or a perfect square. Further, suppose that a_1, a_2, a_3 generate a subgroup of rank 3 in $\mathbb{Q}^*$. Then, one of a_1, a_2, a_3 is a primitive root modulo p for infinitely many primes p.*

We will prove below a result slightly weaker than the one obtained in [23]. It will still serve to illustrate how the lower bound sieve can be applied to such problems.

The strategy is simple enough. By Euler's criterion, a is a primitive root modulo p if and only if

$$a^{(p-1)/q} \not\equiv 1 (\mathrm{mod}\ p), \quad \forall q|p-1.$$

Thus, to ensure that a is a primitive root modulo p, we must verify each of these conditions for every prime divisor q of $p-1$.

Observe that 2 is a primitive root modulo p whenever $p = 4q + 1$ with q prime (see the exercises below). At present, it is unknown if there are infinitely many such primes, the problem being as difficult as the twin prime question.

However, the lower bound sieve and, more precisely, Theorem 10.3.2, guarantees that there are at least

$$\gg \frac{x \log\log x}{(\log x)^3}$$

primes $p \le x$ such that $(p-1)/2$ has all distinct prime factors each greater than

$$x^{1/6}/(\log x)^{2B}.$$

Lemma 10.4.2 *Let Γ be a subgroup of $\mathbb{Q}^*$ of rank r, generated by $a_1, \dots, a_r$ (say). Then the number of primes $p \le x$ for which Γ has good reduction modulo p (that is, we avoid primes dividing the denominators of $a_1, \dots, a_r$) and for which*

$$\#\Gamma(\mathrm{mod}\ p) \le y$$

is $O(y^{1+1/r})$.

Proof Clearly, there are only finitely many primes p for which Γ has bad reduction modulo p. We exclude these primes. Consider the set

$$S := \{a_1^{n_1} \dots a_r^{n_r} : 0 \le n_i \le y^{1/r},\ 1 \le i \le r\}.$$

As the $a_1, \dots, a_r$ are multiplicatively independent (because Γ has rank r), the number of elements of S exceeds

$$\left([y^{1/r}]+1\right)^r > y.$$

Thus, if p is a prime such that

$$\#\Gamma(\mathrm{mod}\ p) \le y,$$

then two distinct elements of S are congruent modulo p. Hence p divides the numerator N of

$$a_1^{m_1} \dots a_r^{m_r} - 1$$

for some $m_1, \dots, m_r$, and the number of such primes is clearly bounded by

$$\log N \le y^{1/r} \sum_{i=1}^{r} \log H(a_i) = O(y^{1/r}),$$

where $H(a_i) := \max(|b_i|, |c_i|)$ and $a_i = b_i/c_i$ with b_i, c_i being coprime integers. Taking into account the number of possibilities for $m_1, \ldots, m_r$, the total number of primes p cannot exceed

$$O(y^{1+1/r}).$$

□

Now suppose $a_1, \ldots, a_6$ are mutually coprime integers. We want to show that there is a positive constant A and a finite set of the form

$$S = \{a_1^{n_1} \ldots a_6^{n_6} : 0 \le n_i \le A\}$$

such that, for some $a \in S$, there are infinitely many primes p for which a is a primitive root modulo p.

Indeed, let p be a prime enumerated in Theorem 10.3.2 and Γ the subgroup generated by $a_1, \ldots, a_6$. By the remark following Theorem 10.3.2, we can ensure that

$$a_i^{(p-1)/2} \equiv -1 \pmod{p}$$

for $1 \le i \le 6$. In particular, the index $[\mathbb{F}_p^* : \Gamma(\mathrm{mod}\ p)]$ is odd. As $(p-1)/2$ has all its prime divisors

$$> x^{1/6}/(\log x)^{2B},$$

we see that if the index $[\mathbb{F}_p^* : \Gamma(\mathrm{mod}\ p)]$ is greater than 1, then

$$\#\Gamma(\mathrm{mod}\ p) \le x^{5/6}(\log x)^{2B}.$$

By Lemma 10.4.2, the number of such primes is

$$O(x^{35/36}(\log x)^{7B/3}),$$

which is negligible compared to the number of primes supplied by Theorem 10.3.2. Thus, we may assume that the index $[\mathbb{F}_p^* : \Gamma(\mathrm{mod}\ p)] = 1$ for at least

$$\gg \frac{x \log \log x}{(\log x)^3}$$

primes $p \le x$.

Now fix such a prime p. Suppose that none of the elements of S are primitive roots modulo p. Choose a primitive root g modulo p and write

$$a_i = g^{e_i} \pmod{p}.$$

Thus, for every 6-tuple $(n_1, \dots, n_6)$ enumerated by S we have

$$\gcd\left(\sum_{i=1}^{6} n_i e_i, p-1\right) > 1.$$

The total number of elements of S is $(A+1)^6$. We remove from this set any tuple $(n_1, \dots, n_6)$ for which

$$\gcd\left(\sum_{i=1}^{6} n_i e_i, p-1\right) \neq 1.$$

As $(p-1)$ has a bounded number of prime factors, there are not too many tuples to eliminate. For instance, the number of tuples with

$$\sum_{i=1}^{6} n_i e_i \equiv 0 (\mathrm{mod}\ 2)$$

is at most $(A+1)^6/2$ (as each of the e_is is odd). For each of the remaining possibilities we may choose $n_1, \dots, n_5$ arbitrarily and then solve for n_6 from

$$n_6 e_6 \equiv -\sum_{i='1}^{5} n_i e_i (\mathrm{mod}\ q)$$

for the prime $q | p-1$ with $q > x^{1/6}/(\log x)^{2B}$. As A is fixed, there can be at most one solution for n_6 in this congruence for x sufficiently large. Thus, if

$$(A+1)^6 - \frac{1}{2}(A+1)^6 - 6(A+1)^5 > 0,$$

we deduce that there is a tuple $(n_1, \dots, n_6)$ in S such that

$$a_1^{n_1} \dots a_6^{n_6}$$

is a primitive root modulo p for the primes enumerated above. Thus, if $A > 11$, we are ensured a solution. This proves:

Theorem 10.4.3 *Let $a_1, \dots, a_6$ be mutually coprime integers greater than 1. Then there are infinitely many primes p for which some member of the finite set*

$$S = \{a_1^{n_1} \dots a_6^{n_6} : 0 \leq n_i \leq 12\}$$

is a primitive root modulo p for infinitely many primes p. Moreover, the number of such primes $p \leq x$ is

$$\gg \frac{x \log \log x}{(\log x)^3}.$$

As indicated in [23], one can obtain stronger results if we apply the Rosser–Iwaniec sieve [34]. By that technique, one can obtain

$$\gg \frac{x}{\log^2 x}$$

primes $p \le x$ with the property that all prime divisors of $(p-1)/2$ are greater than

$$x^{1/4}/(\log x)^B.$$

There are other applications of the lower bound sieve that are too numerous to enumerate here. We will content ourselves by relating one application to the Euclidean algorithm.

An integral domain R is said to be **Euclidean** if there is a map

$$\phi : R\backslash\{0\} \to \mathbb{N}\cup\{0\}$$

such that for any $a, b \in R$, $b \ne 0$, there are $q, r \in R$ such that

$$a = bq + r,$$

with $r = 0$ or $\phi(r) < \phi(b)$.

Any such Euclidean domain is necessarily a principal ideal domain (PID). In 1973, Weinberger [72] proved the remarkable theorem that if K is an algebraic number field with infinitely many units, then the ring of integers of K is Euclidean if and only if it is a PID, provided the generalized Riemann hypothesis holds. Naturally, it would be good to eliminate the use of this hypothesis from this result. In this direction, M. Harper and R. Murty [29] showed that whenever the unit rank of K is greater than 3, the generalized Riemann hypothesis can be dispensed with. A key tool in this work is the lower bound sieve and the large sieve method. We refer the reader to [29] for further details and applications.

10.5 Exercises

1. Prove Lemma 10.2.2.

2. Prove Lemma 10.2.3.

3. Let $d(\cdot)$ be the divisor function and let k be a positive integer. Show that

$$\sum_{m\le x} \frac{d^k(m)}{m} \ll (\log x)^{2^k}.$$

4. Let x, ε be positive real numbers and let

$$y := x^{\alpha+\varepsilon}, \quad z := x^{\beta-\varepsilon}$$

for some $\alpha, \beta > 0$. Show that the conditions

$$yz^2 \le x^{\frac{1}{2}-\varepsilon}$$

and

$$\lim_{x \to \infty} \left(1 - \log\left(\frac{\log y}{\log z} \right) \right) > 0$$

are satisfied if and only if

$$\alpha = \frac{1}{4}, \beta = \frac{1}{8}.$$

5. Show that 2 is a primitive root modulo p if p is of the form $4q+1$ with q a prime.

6. In the proof of Theorem 10.4.3, what goes wrong if we take five mutually coprime integers greater than 1?

7. Fix a prime $q < x^\theta$ with $\theta < 1/2$. Show that the number of primes $p \equiv 1 \pmod{q}$ such that $(p-1)/2q$ has all its prime factors $> x^\delta$ for some $\delta > 0$ is

$$\ll \frac{x}{q(\log x)^2}.$$

8. Assuming the result stated at the end of Section 10.4, deduce that there are

$$\gg \frac{x}{(\log x)^2}$$

primes $p \le x$ with the property that all the prime divisors of $(p-1)/2$ are greater than

$$x^{1/4}(\log x)^B.$$

Deduce that if a, b, c are three distinct prime numbers, then one of them is a primitive root modulo p for infinitely many primes p.

9. Assuming the Elliott–Halberstam conjecture (see Chapter 9), show that for any distinct prime numbers a, b, at least one of them is a primitive root modulo p.

11

New directions in sieve theory

Nearly a century after the discovery of Brun's sieve, we can look back and see how the subject has developed, and, to some extent, indicate how it may develop in the next 100 years.

One of the dominant themes of the twentieth century number theory has been the 'modular connection'. In 1955, Yutaka Taniyama (1927–58) first hinted at a connection between elliptic curves and automorphic forms. The Langlands program has absorbed this theme and the connection is expected to hold in a wider context. At the heart of the Langlands program lies the 'Rankin–Selberg method', which signals an 'orthogonality principle' for automorphic representations on $GL(n)$. This point of view has suggested one mode of generalizing the large sieve inequalities of analytic number theory.

In a series of remarkable papers, H. Iwaniec and his school have developed the 'modular connection' and the cognate 'spectral connection' as it applies to $GL(2)$ analogues of the large sieve inequality (see, for example, [2,3,9,33]).

In this chapter, we give a brief overview of the work of Duke and Kowalski [10] that suggests a future direction for the large sieve method. No doubt, there will be other directions of development, but the authors do not have a crystal ball to perceive them.

11.1 A duality principle

The large sieve inequality can be reduced to a statement in linear algebra, which in turn can be proven using matrix theory. This is the point of view of Elliott [11, pp. 150–70]. We review this below, as it is one of the key ingredients in 'non-abelian' generalizations of the large sieve, initiated (and, to a large extent, developed) by Iwaniec [33].

201

Theorem 11.1.1   *(A principle of duality)*
Let $c_{ij}, 1 \leq i \leq m, 1 \leq j \leq n$, be mn complex numbers. Let λ be a non-negative real number. Then the inequality

$$\sum_{1 \leq i \leq m} \left| \sum_{1 \leq j \leq n} c_{ij} a_j \right|^2 \leq \lambda \sum_{1 \leq j \leq n} |a_j|^2$$

holds for any complex numbers $a_1, \ldots, a_n$ if and only if the inequality

$$\sum_{1 \leq j \leq n} \left| \sum_{1 \leq i \leq m} c_{ij} b_i \right|^2 \leq \lambda \sum_{1 \leq i \leq m} |b_i|^2$$

holds for any complex numbers $b_1, \ldots, b_m$.

Proof   First, let us recall some standard notation. For a vector v with complex components, v^t denotes its transpose, $||v||$ denotes its length in the Euclidean space, and $\bar{v}$ denotes its complex conjugate.

Now, for $a_1, \ldots, a_n \in \mathbb{C}$ and $b_1, \ldots, b_m \in \mathbb{C}$, let $\underline{a}$ and $\underline{b}$ denote the column vectors $(a_1, \ldots, a_n)^t$ and $(b_1, \ldots, b_m)^t$, respectively. We consider the matrix

$$C := (c_{ij})_{1 \leq i \leq m, 1 \leq j \leq n},$$

and we observe that the first inequality of the theorem can be rewritten as

$$||C\underline{a}||^2 \leq \lambda ||\underline{a}||^2 \quad \forall \underline{a}$$

and the second one can be rewritten as

$$||\underline{b}^t C||^2 \leq \lambda ||\underline{b}||^2 \quad \forall \underline{b}.$$

Let us assume that the first inequality holds. By the Cauchy–Schwarz inequality, we have

$$||\underline{b}^t C\underline{a}||^2 \leq ||\underline{b}^t||^2 ||C\underline{a}||^2 \leq \lambda ||\underline{b}||^2 ||\underline{a}||^2 \quad \forall \underline{a},$$

the last inequality following from our assumption. Now set

$$\underline{a} := \overline{C}^t \overline{\underline{b}},$$

so that

$$||\underline{b}^t C||^4 \leq \lambda ||\underline{b}||^2 ||\underline{b}^t C||^2.$$

From this we deduce that

$$||\underline{b}^t C||^2 \leq \lambda ||\underline{b}||^2 \quad \forall \underline{b},$$

as required.

The converse implication is immediate upon interchanging the roles of i and j. $\square$

More can be said about λ appearing in the previous theorem. The matrix $A = \overline{C}^t C$ is an $n \times n$ Hermitian matrix. Consequently, all of its eigenvalues, denoted $\lambda_1, \dots, \lambda_n$, are real and we may diagonalize A by a unitary transformation U. More precisely, there is a matrix U satisfying

$$U\overline{U}^t = I$$

and

$$\|C\underline{x}\|^2 = \underline{\overline{x}}^t A \underline{x} = \underline{\overline{y}}^t \overline{U}^t \overline{C}^t CU\underline{y} = \sum_{1 \le j \le n} \lambda_j |y_j|^2,$$

where $\underline{x} := U\underline{y}$. Without loss of generality we may assume that

$$\lambda_1 \ge \lambda_2 \ge \dots \ge \lambda_n \ge 0,$$

so that

$$\|C\underline{x}\|^2 \le \lambda_1 \|\underline{y}\|^2 = \lambda_1 \|\underline{x}\|^2,$$

as U is a unitary transformation (and hence preserves lengths). Thus λ in the theorem can be taken to be the maximal eigenvalue of $\overline{C}^t C$, and if the a_j's are unrestricted, this is the best choice possible.

By duality, we can consider $B = C\overline{C}^t$ instead of $\overline{C}^t C$. As B is Hermitian, its eigenvalues are real and, again, we may order them as

$$\mu_1 \ge \mu_2 \ge \dots \ge \mu_n \ge 0$$

to deduce that the inequality of the theorem holds with $\lambda = \mu_1$, which is best possible. Thus the theorem implies $\lambda_1 = \mu_1$.

In fact, more is true. One can show that $\lambda_i = \mu_i$ for $i \le \min(m, n)$ and that the remaining eigenvalues are all zero (see [11, pp. 163–4]).

The large sieve inequality (Theorem 8.7) can be derived from the following more general inequality. Let x_j, $1 \le j \le R$, be real numbers that satisfy

$$\|x_j - x_k\| \ge \delta > 0 \ (j \ne k),$$

where $\|y\|$ denotes now the distance of y to the nearest integer. That is,

$$\|y\| := \min (y - [y], [y] + 1 - y)$$

(we hope that the reader will not confuse it with the length of a vector). Then the large sieve inequality can be deduced from (see Exercise 2)

$$\sum_{1 \le j \le R} \left| \sum_{1 \le n \le N} a_n e^{2\pi i n x_j} \right|^2 \le (N + \delta^{-1}) \sum_{1 \le n \le N} |a_n|^2.$$

This last inequality can be proven by determining the largest eigenvalue of the $R \times R$ matrix

$$C\overline{C}^t,$$

where

$$C := \left(e^{2\pi i x_j n}\right)_{1 \leq j \leq R, 1 \leq n \leq N}.$$

Notice that the (j, k)-th entry of $C\overline{C}^t$ is

$$\sum_{1 \leq n \leq N} e^{2\pi i (x_j - x_k) n},$$

which is a geometric series that can be summed directly. If $x_j \neq x_k$, this sum is easily seen to be

$$2i e^{\pi i (x_j - x_k)} \left(\frac{e^{2\pi i (x_j - x_k) N} - 1}{\sin \pi (x_j - x_k)} \right).$$

By Exercise 3, this is

$$\ll \|x_j - x_k\|^{-1}.$$

Now, given any $n \times n$ matrix $A = (a_{ij})$, it is easy to see (Exercise 6) that all the eigenvalues λ of A lie in the discs

$$|\lambda - a_{ii}| \leq \sum_{j \neq i} |a_{ij}|.$$

Since $\|x_j - x_k\| \geq \delta$ for any $j \neq k$, a little reflection (Exercise 5) shows that

$$\sum_{j \neq k} \|x_j - x_k\|^{-1} \ll \delta^{-1} \log \frac{1}{\delta},$$

which gives a slightly weaker form of the large sieve inequality. A refined analysis (see [11, pp. 166–170]) gives the sharper inequality stated above.

As discussed in Theorem 8.3.1, the large sieve inequality must be reformulated in terms of Dirichlet characters before it can be applied to deduce important consequences to the theory of L-functions or the Bombieri–Vinogradov theorem (Theorem 9.2.1). More precisely, the form of the large sieve is the inequality

$$\sum_{d \leq z} \frac{d}{\phi(d)} \sum_{\chi}^* \left| \sum_{n \leq x} a_n \chi(n) \right|^2 \leq (z^2 + 4\pi x) \sum_{n \leq x} |a_n|^2,$$

where the summation $\sum_{\chi}^*$ is over primitive Dirichlet characters modulo d. The transition is effected by the orthogonality of the Dirichlet characters and the inequality should be viewed as some form of 'quasi'-orthogonality. This viewpoint suggests the study of sums of the form

$$\sum_f \left| \sum_n a_n \chi_f(n) \right|^2 ,$$

where $\chi_f(n)$ is to be thought of as a generalized character and where f ranges over a suitable family. We formalize this in the next section.

11.2 A general formalism

The following set-up is suggested by a paper of Duke and Kowalski [10]. Let $\mathcal{F}$ be a finite set. For each $f \in \mathcal{F}$, suppose that we are given a sequence

$$\lambda_f(n), \quad n \geq 1.$$

Let $(a_n)_{n \geq 1}$ be a sequence of complex numbers and x a positive real number. The goal is to derive an inequality of the form

$$\sum_{f \in \mathcal{F}} \left| \sum_{n \leq x} a_n \lambda_f(n) \right|^2 \leq \Delta \sum_{n \leq x} |a_n|^2$$

for some $\Delta > 0$.

We will suppose that the elements of $\mathcal{F}$ form a 'quasi'-orthogonal family in the following sense: there exist non-negative numbers α, β such that, for any $f, g \in \mathcal{F}$,

$$\sum_{n \leq x} \lambda_f(n) \overline{\lambda_g(n)} = c(f, g)x + O\left(x^\alpha (\#\mathcal{F})^\beta\right), \qquad (11.1)$$

where the implied constant is absolute and where $c(f, g) = 0$ unless $f = g$, in which case we suppose that

$$c(f, f) = O\left((\#\mathcal{F})^\varepsilon\right)$$

for any $\varepsilon > 0$.

By duality we need to estimate

$$\sum_{n \leq x} \left| \sum_{f \in \mathcal{F}} b_f \lambda_f(n) \right|^2$$

for any sequence $(b_f)_{f \in \mathcal{F}}$ of complex numbers. Expanding and using that $2|b_f b_g| \le |b_f|^2 + |b_g|^2$, we get by (11.1) that

$$\sum_{f,g} b_f \overline{b_g} \sum_{n \le x} \lambda_f(n) \overline{\lambda_g(n)} = x \sum_f |b_f|^2 c(f,f) + O\left(x^{\alpha}(\#\mathcal{F})^{\beta+1}\right) \sum_f |b_f|^2$$

$$\ll \left(x(\#\mathcal{F})^{\varepsilon} + O\left(x^{\alpha}(\#\mathcal{F})^{\beta+1}\right)\right) \sum_f |b_f|^2.$$

This proves:

Theorem 11.2.1 *Let $\mathcal{F}$ be a finite set and for each $f \in \mathcal{F}$, let $(\lambda_f(n))_{n \ge 1}$ be a sequence of complex numbers. Let $(a_n)_{n \ge 1}$ be a sequence of complex numbers. Let $x, \varepsilon > 0$. Under assumption (11.1) we have*

$$\sum_{f \in \mathcal{F}} \left| \sum_{n \le x} a_n \lambda_f(n) \right|^2 \ll \left(x(\#\mathcal{F})^{\varepsilon} + O\left(x^{\alpha}(\#\mathcal{F})^{\beta+1}\right)\right) \sum_{n \le x} |a_n|^2.$$

Let us note that the Cauchy–Schwarz inequality gives the estimate

$$\left(x(\#\mathcal{F})^{1+\varepsilon} + O\left(x^{\alpha}(\#\mathcal{F})^{\beta+1}\right)\right) \sum_{n \le x} |a_n|^2$$

(see Exercise 7). We see that Theorem 11.2.1 provides a saving of $\#\mathcal{F}$ in the main term.

To apply this theorem, we must have a ready supply of families $\mathcal{F}$ that satisfy (11.1). A natural family is provided by normalized Hecke eigenforms in the theory of modular forms. Indeed, if f and g are cusp forms of weights k_1 and k_2, respectively, with Fourier expansions

$$f(z) = \sum_{n \ge 1} \lambda_f(n) n^{\frac{k_1-1}{2}} e^{2\pi i n z},$$

$$g(z) = \sum_{n \ge 1} \lambda_g(n) n^{\frac{k_2-1}{2}} e^{2\pi i n z},$$

then the Rankin–Selberg L-series

$$\sum_{n \ge 1} \frac{\lambda_f(n) \overline{\lambda_g(n)}}{n^s}$$

has been studied (independently) by Rankin [56] and Selberg [57]. Their results imply that if we take $\mathcal{F}$ to be the set of normalized Hecke eigenforms of level $\le Q$, then (11.1) holds with $\alpha = 1/2$, $\beta = 1/2$ and with $c(f, g)$ equal to zero unless $f = g$, in which case $c(f, f)$ is related to the Petersson inner product (f, f). One also has the estimate on $c(f, f)$ required by (11.1).

This observation opens up a wide spectrum of examples and potential applications. Relation (11.1) holds in a wider context of Rankin–Selberg theory in the Langlands program. We refer the reader to [6] for a general introduction to the theory.

11.3 Linnik's problem for elliptic curves

In 1941, when Linnik [36] introduced his large sieve, he was motivated by the following question: let d be a natural number and χ a primitive Dirichlet character modulo d; what is the size of the smallest $n_\chi = n$ such that $\chi(n) \neq 1$?

In the special case that χ is the quadratic character (that is, $\chi^2 = \chi_0$, the trivial character) and $d =: q$ is a prime, this is the question concerning the size of the least quadratic non-residue modulo q and one has the celebrated conjecture of Vinogradov that $n_\chi \ll q^\varepsilon$ for any $\varepsilon > 0$. The generalized Riemann hypothesis (see [41, Chapter 13]) implies

$$n_\chi \ll (\log q)^2.$$

Using his large sieve, Linnik obtained a statistical result towards this question. More precisely, if $D > 0$ and $\alpha > 1$, and if we let

$$N(D, \alpha)$$

be the number of primitive characters χ of modulus $d \leq D$ such that $\chi(n) = 1$ for all $n \leq (\log D)^\alpha$, then Linnik's argument shows that

$$N(D, \alpha) \ll D^{\frac{2}{\alpha} + \varepsilon}$$

for any $\varepsilon > 0$. Since the total number of characters under consideration is about D^2, this result says that the number of characters χ with $n_\chi > (\log D)^\alpha$ is $o(D^2)$ for any $\alpha > 2$. In particular, one can conclude that Vinogradov's hypothesis is 'almost always' true in the probabilistic sense described above.

After the advent of the Langlands program and, to some extent, the important work of Iwaniec [33], automorphic representations are viewed as higher dimensional analogues of Dirichlet characters. The discussion in the previous section and its brief allusion to Rankin–Selberg theory justifies, to some extent, this viewpoint. Thus it has been a useful background theme to investigate to what extent theorems concerning Dirichlet characters can be generalized to the higher dimensional 'automorphic' context. Poetic as this may sound, the transfer of ideas is not easy and our understanding even in the GL(2) context has been meagre. However, some results can be obtained, suggesting that 'non-abelian' analogues of the large sieve inequality do exist.

This perspective first arose in the fundamental work of Iwaniec [33] and then was developed in the foundational paper [9]. These works focused on the GL(2) analogues. In [10], Duke and Kowalski presented a general approach on how to generalize the large sieve inequality to GL(n). It is this presentation that we will follow here.

Inspired by Linnik's problem, Duke and Kowalski consider the following problem: given two non-isogenous elliptic curves E_1 and E_2 over $\mathbb{Q}$, how large can n be such that E_1 and E_2 have the same number of points modulo p for all primes $p \leq n$ where E_1 and E_2 have good reduction? We can phrase this question in another way, as follows. Given an elliptic curve E over $\mathbb{Q}$, let $a_p(E)$ be defined by

$$a_p(E) := p + 1 - \#E(\mathbb{F}_p)$$

for primes p of good reduction. It is known that two elliptic curves E_1 and E_2 are isogenous over $\mathbb{Q}$ if and only if $a_p(E_1) = a_p(E_2)$ for all but finitely many primes p. Thus, given two non-isogenous elliptic curves E_1 and E_2 over $\mathbb{Q}$, the problem is to find the size of the smallest prime p so that $a_p(E_1) \neq a_p(E_2)$. We shall refer to it as **Linnik's problem for elliptic curves**.

This question comes up in many contexts and was first discussed by Serre [64] in his determination of the Galois groups obtained by adjoining the ℓ-division points of a given elliptic curve to $\mathbb{Q}$. He proved, under the generalized Riemann hypothesis (for Dedekind zeta functions), that one can find such a prime p satisfying

$$p \ll (\log D)^2,$$

where $D := \max(N_1, N_2)$ with N_1, N_2 equal to the conductors of E_1, E_2, respectively. Thus, a natural question to ask is if one can prove a result analogous to Linnik's without any hypothesis in the elliptic curve context. This is what is proven in [10]:

Theorem 11.3.1 *(Duke and Kowalski)*
Let $D > 0, \alpha > 1$. Let

$$M(D, \alpha)$$

be the maximal number of isogeny classes of semistable elliptic curves over $\mathbb{Q}$, with conductor $\leq D$, such that for every prime $p \leq (\log D)^\alpha$ they have the same number of points modulo p. Then, for any $\varepsilon > 0$,

$$M(D, \alpha) \ll D^{\frac{10}{\alpha} + \varepsilon}.$$

Fouvry *et al.* [19] have shown that the number of semistable elliptic curves of conductor $\leq D$ is

$$\gg D^{\frac{5}{6}}.$$

That this is the expected result can be seen, roughly, as follows. Given an elliptic curve $E : y^2 = x^3 + ax + b$ over $\mathbb{Q}$, the discriminant of the equation is $4a^3 + 27b^2$ (which may not be equal to the discriminant of the curve!). The number of choices for a and b so that $4a^3 + 27b^2 \leq D$ and $4a^3 + 27b^2$ is squarefree is, approximately,

$$\gg D^{1/3} D^{1/2} = D^{5/6}.$$

This heuristic argument shows that the result of [19] is of the correct order of expected magnitude.

Thus, if $10/\alpha < 5/6$, that is, if $\alpha > 12$, we obtain from Theorem 11.3.1 a statistical result of Linnik type in the elliptic curve context.

In view of the spectacular work of Wiles [73] and his school, all elliptic curves over $\mathbb{Q}$ are modular. That is, to each elliptic curve E over $\mathbb{Q}$ we can associate a weight 2 cusp form f_E such that $a_p(E) = \lambda_{f_E}(p)$, where $\lambda_{f_E}(p)$ denotes the p-th Fourier coefficient of f_E. Hence the Linnik problem for elliptic curves can be rephrased in terms of cups forms. This is the strategy adopted in [10]. We therefore consider the larger class of cusp forms of level $\leq Q$, derive a 'large sieve'-type inequality for them, and then deduce Theorem 11.3.1.

11.4 Linnik's problem for cusp forms

Let k be a positive even integer. For a positive integer d, let

$$S_k(d)^+$$

be the set of primitive cusp forms of weight k and level d. We know that the elements of $S_k(d)^+$ are Hecke eigenforms.

Let $D > 0$ and let

$$S_k(\leq D)^+$$

denote the set of primitive cusp forms of weight k and level $\leq D$. For $f \in S_k(\leq D)^+$ we write

$$f(z) = \sum_{n \geq 1} \lambda_f(n) n^{\frac{k-1}{2}} e^{2\pi i n z}$$

for its Fourier expansion at $i\infty$.

There are precise formulae for the cardinality of $S_k(d)^+$ and we can deduce that

$$\#S_k(\leq D)^+ \sim c(k)D^2$$

for some positive constant $c(k)$ (see [66, pp. 25–46] for details).

We will be interested in a special subset of $S_k(\leq D)^+$, denoted

$$S_k(\leq D)^\#,$$

consisting of 'non-monomial' forms. One can characterize the 'monomial' forms as those f for which there exists a quadratic Dirichlet character χ such that, for all primes p, we have

$$\lambda_f(p) = \lambda_f(p)\chi(p).$$

These are the so-called 'forms of CM-type'. One can show that the collection of monomial elements of $S_k(\leq D)^+$ is of size $o\left(D^2\right)$, so that

$$\#S_k(\leq D)^\# \sim c(k)D^2$$

as $D \to \infty$ (in fact, one can show that the number of monomial forms is $O\left(D^{1+\varepsilon}\right)$ for any $\varepsilon > 0$).

The main theorem in [10] is:

Theorem 11.4.1 *Let k be a positive even integer, $D, \beta, \varepsilon > 0$. Let $(a_n)_{n\geq1}$ be a sequence of complex numbers.*

1. If $\beta > 6$, then

$$\sum_{f \in S_k(\leq D)^\#} \left| \sum_{n \leq D^\beta} a_n \lambda_f(n) \right|^2 \ll D^{\beta+\varepsilon} \sum_{n \leq D^\beta} |a_n|^2.$$

2. If $\beta > 10$, then

$$\sum_{f \in S_k(\leq D)^\#}^* \left| \sum_{n \leq D^\beta} \mu^2(n) a_n \lambda_f\left(n^2\right) \right|^2 \ll D^{\beta+\varepsilon} \sum_{n \leq D^\beta} \mu^2(n)|a_n|^2,$$

where the star on the summation indicates that we sum over those f 'up to quadratic equivalence'. That is, two forms $f, g \in S_k(\leq D)^\#$ are said to be quadratic equivalent if there exists a quadratic character χ such that, for all p, $\lambda_f(p) = \lambda_g(p)\chi(p)$. The summation is over representatives of the equivalence classes.

We will discuss the proof of this theorem in the next section. In this section we will indicate how one may deduce Theorem 11.3.1 from it.

Proof of Theorem 11.3.1 Fix k, D, β, ε and $(a_n)_{n \geq 1}$ as in Theorem 11.4.1. Also, fix $\alpha > 1$ and a set $\mathcal{P}$ of primes, of natural density δ. Thus

$$\#\{p \in \mathcal{P} : p \leq x\} \sim \delta \pi(x)$$

as $x \to \infty$.

We will say that two elements $f, g \in S_k(\leq D)^\#$ are *equivalent*, denoted $f \sim g$, if

$$\lambda_f(p) = \lambda_g(p) \ \forall p \in \mathcal{P}, \ p \leq (\log D)^\alpha.$$

Let us fix $f_0 \in S_k(\leq D)^\#$ and let d_0 be the level of f_0. We introduce the notation

$$\mathcal{P}(D) := \{p \leq (\log D)^\alpha : p \nmid d_0\},$$

$$\mathcal{P}_1(D) := \{p \in \mathcal{P}(Q) : |\lambda_{f_0}(p)| \geq 1/2\},$$

$$\mathcal{P}_2(D) := \{p \in \mathcal{P}(Q) : |\lambda_{f_0}(p^2)| \geq 1/2\}.$$

Since d_0 has $\ll \log D$ prime divisors, the set $\mathcal{P}(D)$ satisfies

$$\#\mathcal{P}(D) \sim \frac{\delta(\log D)^\alpha}{\alpha \log \log D},$$

as $\alpha > 1$. For any $p \in \mathcal{P}(D)$ we have

$$\lambda_{f_0}(p)^2 - \lambda_{f_0}(p^2) = 1,$$

so that, for some $i_0 \in \{1, 2\}$,

$$\#\mathcal{P}_{i_0}(D) \geq \#\mathcal{P}(D)/2 \geq \frac{\delta(\log D)^\alpha}{3\alpha \log \log D}.$$

Let m be a positive integer to be chosen soon. We denote by

$$\mathcal{N}(D)$$

the set of squarefree integers n that have m prime factors, all from $\mathcal{P}_{i_0}(D)$. Let N be the largest element of $\mathcal{N}(D)$, so that

$$N \leq (\log D)^{\alpha m} =: N' \text{ (say)}.$$

We assume that m is chosen such that N' is less than (but near) D^β, with β to be chosen so that we may apply Theorem 11.4.1.

We remark that if $f \sim f_0$ in the sense defined above, then $\lambda_f(p) = \lambda_{f_0}(p)$ for all $p \in \mathcal{P}_i(D)$, hence $\lambda_f(n) = \lambda_{f_0}(n)$ if n has all its prime factors in $\mathcal{P}_i(D)$, where $1 \leq i \leq 2$. Moreover, if $n \in \mathcal{N}(D)$, then

$$|\lambda_f(n^{i_0})| = |\lambda_{f_0}(n^{i_0})| \geq 2^{-m}. \tag{11.2}$$

Now let us deduce Theorem 11.3.1. For this, it is enough to take $k = 2$, and in this situation the cusp forms f appearing in the inequalities given by Theorem 11.4.1 will correspond to elliptic curves over $\mathbb{Q}$. We will also consider only those f's belonging to the equivalence class of the fixed form f_0, so that we count elliptic curves in the same isogeny class. We choose

$$a_n := \begin{cases} \lambda_{f_0}\left(n^{i_0}\right) & \text{if } n \in \mathcal{N}(D) \\ 0 & \text{otherwise,} \end{cases}$$

In the case $i_0 = 1$, we use the first inequality of Theorem 11.4.1 to obtain

$$M(D, \alpha) \left| \sum_{n \in \mathcal{N}(D)} |\lambda_{f_0}(n)|^2 \right|^2 \ll D^{\beta+\varepsilon} \sum_{n \in \mathcal{N}(D)} |\lambda_f(n)|^2.$$

Now (11.2) implies that

$$\sum_{n \in \mathcal{N}(D)} |\lambda_{f_0}(n)|^2 \geq 2^{-2m} \# \mathcal{N}(D).$$

Thus

$$M(D, \alpha) \ll \frac{D^{\beta+\varepsilon} 2^{2m}}{\# \mathcal{N}(D)}.$$

In the case $i_0 = 2$, let $M^{(2)}(D, \alpha)$ be the size of $M(D, \alpha)$ up to quadratic twists. Then, using the second inequality of Theorem 11.4.1, we get that, for $\beta > 10$,

$$M^{(2)}(D, \alpha) \left| \sum_{n \in \mathcal{N}(D)} |\lambda_{f_0}\left(n^2\right)|^2 \right|^2 \ll D^{\beta+\varepsilon} \sum_{n \in \mathcal{N}(D)} |\lambda_{f_0}\left(n^2\right)|^2,$$

so that, as before,

$$M^{(2)}(D, \alpha) \ll \frac{D^{\beta+\varepsilon} 2^{2m}}{\# \mathcal{N}(D)}.$$

We choose

$$m := \left\lceil \frac{\beta \log Q}{\alpha \log \log Q} \right\rceil,$$

and so $2^{2m} = O(D^\varepsilon)$. By unique factorization and Stirling's formula,

$$\# \mathcal{N}(D) \geq \binom{\# \mathcal{P}_{i_0}(D)}{m} \gg D^{\beta\left(\frac{\alpha-1}{\alpha}\right)-\varepsilon}.$$

Hence

$$M^{(2)}(D, \alpha) \ll D^{\frac{\beta}{\alpha}+\varepsilon}.$$

Since any equivalence class contributes at most one semistable curve, this completes the proof of Theorem 11.3.1. □

We remark that the number of quadratic twists of a given f that may be in any equivalence class is at most $O\left(D^{\frac{1}{2}+\varepsilon}\right)$, hence this would have to be taken into account in the general case. However, these estimates can be refined and we refer the reader to [10, p. 15] for further details.

The interested reader should also compare this argument with the classical one of Linnik, which can be found in [1, p. 7].

11.5 The large sieve inequality on GL(n)

In this section we adopt the general approach of [10] and outline, very briefly, how one may prove Theorem 11.4.1. For any comprehensive proof the prerequisites are formidable and it would be out of the scope of this book to provide complete details. Our goal is to indicate (rather vaguely) how such a theory is to be developed. As we feel the theory will have profound applications in the future, it would be short-sighted on our part to leave out a discussion of it. For the terminology and background, we refer the reader to [22].

Let π be a cuspidal automorphic representation of $\mathrm{GL}_n(A_\mathbb{Q})$, where $A_\mathbb{Q}$ denotes the adele ring of the rational number field. To each such π, Langlands has attached an L-function:

$$L(s, \pi) := \sum_{n \geq 1} \frac{\lambda_\pi(n)}{n^s}.$$

We fix the infinity component π_∞ of π and the central character η of π. For any integer $d \geq 1$, let

$$\mathrm{Aut}(d)$$

be the set of cuspidal automorphic representations π of $\mathrm{GL}_n(A_\mathbb{Q})$ such that

1. π_∞ is the infinity component of π and η is its central character;
2. π satisfies the Ramanujan conjecture in the sense that

$$\lambda_\pi(n) = O(n^\varepsilon)$$

for any $\varepsilon > 0$;
3. the conductor of π is d.

For $D > 0$, let

$$\mathrm{Aut}(\leq D) := \cup_{d \leq D}\mathrm{Aut}(d).$$

For each d, suppose that we are given a set $S(d) \subseteq \mathrm{Aut}(d)$ and define

$$S(\leq D) := \cup_{d \leq D}S(d).$$

The goal is to prove an inequality of the form

$$\sum_{\pi \in S(\leq D)}\left|\sum_{n \leq N}a_n\lambda_\pi(n)\right|^2 \ll N^{1+\varepsilon}\sum_{n \leq N}|a_n|^2$$

for D and N in a suitable range. As remarked in Section 11.1, this is equivalent, by duality, to

$$\sum_{n \leq N}\left|\sum_{\pi \in S(\leq D)}b_\pi\lambda_\pi(n)\right|^2 \ll N^{1+\varepsilon}\sum_{\pi \in S(\leq D)}|b_\pi|^2.$$

Since we seek only an upper bound, we choose a smooth, positive, compactly supported test function

$$\psi : \mathbb{R}_+ \longrightarrow \mathbb{R}_+$$

so that $\psi(x) = 1$ for $0 \leq x \leq 1$ and $\psi(x) \geq 0$ for all $x \in \mathbb{R}_+$. Then we must estimate

$$\sum_{n \geq 1}\left|\sum_{\pi \in S(\leq D)}b_\pi\lambda_\pi(n)\right|^2\psi\left(\frac{n}{N}\right).$$

Expanding the square and interchanging the order of summation gives

$$\sum_{\pi_1,\pi_2 \in S(\leq D)}b_{\pi_1}\overline{b_{\pi_2}}\sum_{n \geq 1}\lambda_{\pi_1}(n)\overline{\lambda_{\pi_2}(n)}\psi\left(\frac{n}{N}\right).$$

Let us consider the Dirichlet series

$$L_b(s, \pi_1 \otimes \tilde{\pi}_2) := \sum_{n \geq 1}\frac{\lambda_{\pi_1}(n)\overline{\lambda_{\pi_2}(n)}}{n^s},$$

which we call the 'naive' convolution L-series of π_1 and π_2. It is closely related to the Rankin–Selberg convolution $L(s, \pi_1 \otimes \tilde{\pi}_2)$: there exists a Dirichlet series

$$H(s; \pi_1, \tilde{\pi}_2),$$

which converges absolutely for $\mathrm{Re}(s) > 1/2$, so that

$$L_b(s, \pi_1 \otimes \tilde{\pi}_2) = H(s; \pi_1, \tilde{\pi}_2)L(s, \pi_1 \otimes \tilde{\pi}_2).$$

Moreover, we have, for any $\varepsilon > 0$ and uniformly for $\mathrm{Re}(s) = \sigma > 1/2$, a bound

$$H(s; \pi_1, \tilde{\pi}_2) \ll [q_1, q_2]^\varepsilon H(\sigma),$$

where H is a fixed Dirichlet series absolutely convergent in $\mathrm{Re}(s) > 1/2$ and satisfying

$$H(\sigma) \ll \left(\sigma - \frac{1}{2}\right)^{-A}$$

for some $A > 0$; here, $[q_1, q_2]$ denotes the least common multiple of the conductors q_1, q_2 of π_1, π_2, respectively.

Define

$$\hat{\psi}(s) := \int_0^\infty \psi(x) x^s \frac{\mathrm{d}x}{x},$$

so that

$$\psi(x) = \frac{1}{2\pi i} \int_{c-i\infty}^{c+i\infty} \hat{\psi}(s) x^{-s} \, \mathrm{d}s$$

for $c > 2$. Hence

$$\sum_{n \geq 1} \lambda_{\pi_1}(n) \overline{\lambda_{\pi_2}}(n) \psi\left(\frac{n}{N}\right) = \frac{1}{2\pi i} \int_{c-i\infty}^{c+i\infty} N^s \hat{\psi}(s) H(s; \pi_1, \tilde{\pi}_2) L(s, \pi_1 \otimes \tilde{\pi}_2) \, \mathrm{d}s.$$

Now we can apply standard techniques of analytic number theory (see [45, Chapter 4]). We move the line of integration to $\mathrm{Re}(s) = \frac{1}{2} + c$, with $c < 1/2$ to be chosen later. The Mellin transform $\hat{\psi}$ is easily seen to be holomorphic for $\mathrm{Re}(s) > 0$ and rapidly decreasing in any vertical strip $0 < \delta < \mathrm{Re}(s) < b$ for every positive b. By the Phragmén–Lindelöf principle (see [45, Chapter 8]),

$$\left| L\left(\frac{1}{2} + c + it, \pi_1 \otimes \tilde{\pi}_2\right) \right| \ll D^{2n\left(\frac{1}{4} - \frac{c}{2}\right)} |t|^E$$

for some E, using the work of Bushnell–Henniart on conductors of tensor products of two automorphic representations. The final result is that the sum in question is

$$= \delta(\pi_1, \pi_2) \hat{\psi}(1) N R_{\pi_1} + O\left(N^{\frac{1}{2} + \frac{n}{2\beta} + \varepsilon}\right)$$

when $N > D^\beta$, where $\delta(\cdot, \cdot)$ is the Kronecker delta function and R_{π_1} is the residue of $L(s, \pi_1 \otimes \tilde{\pi}_1)$ at $s = 1$. Moreover, R_{π_1} can be estimated to be $O(D^\varepsilon)$ by using the Ramanujan bound for $\lambda_{\pi_1}(n)$ that we have assumed.

If we now assume that, for some a,

$$\#S(\leq D) \ll D^a,$$

then putting everything together gives

$$\sum_{\pi \in S(\leq D)} \left| \sum_{n \leq N} a_n \lambda_\pi(n) \right|^2 \ll N^{1+\varepsilon} \sum_{n \leq N} |a_n|^2,$$

provided $N > D^\beta$ and $\beta > 2a + n$.

This completes our discussion of the theorem.

11.6 Exercises

1. Let C be an $m \times n$ matrix. Show that the non-zero eigenvalues of $\overline{C}^t C$ and $C\overline{C}^t$ are equal.

2. Deduce Theorem 8.7 from

$$\sum_{1 \leq j \leq R} \left| \sum_{1 \leq n \leq N} a_n e^{2\pi i n x_j} \right|^2 \leq \left(N + \delta^{-1} \right) \sum_{1 \leq n \leq N} |a_n|^2,$$

where $||x_j - x_k|| \geq \delta$ for $j \neq k$.

3. Let $C = (c_{ij})$ be an $m \times n$ matrix with complex entries. Let λ be the largest eigenvalue of the Hermitian matrix $\overline{C}^t C$. Show that

$$n\lambda \geq \sum_{i=1}^{m} \sum_{j=1}^{n} |c_{ij}|^2.$$

By repeating the argument for $C\overline{C}^t$, deduce that

$$\lambda \min(m, n) \geq \sum_{i=1}^{m} \sum_{j=1}^{n} |c_{ij}|^2.$$

4. Infer from the preceding exercises that, in the large sieve inequality, the factor $z^2 + 4\pi x$ cannot be replaced by a quantity smaller than

$$\max(x, \sum_{d \leq z} \phi(d)).$$

5. Prove that for some constant $c > 0$,

$$|\sin x| \geq c||x||.$$

6. Let $A = (a_{ij})$ be an $n \times n$ matrix. Show that any eigenvalue λ of A satisfies

$$|\lambda - a_{ii}| \leq \sum_{j \neq i} |a_{ij}|, \ 1 \leq i \leq n.$$

7. Show that if (11.1) holds, then

$$\sum_{f \in \mathcal{F}} \left| \sum_{n \leq x} a_n \lambda_f(n) \right|^2 \ll \left(x (\#\mathcal{F})^{1+\varepsilon} + O\left(x^\alpha (\#\mathcal{F})^{\beta+1} \right) \right) \sum_{n \leq x} |a_n|^2.$$

8. By applying the duality principle to the Turán–Kubilius inequality (see the exercises in Chapter 3), deduce that

$$\sum_{p^k \leq x} p^k \left| \sum_{n \leq x, p^k \| n} a_n - p^{-k} \sum_{n \leq x} a_n \right|^2 \ll x \sum_{n \leq x} |a_n|^2,$$

uniformly for all real x and all complex numbers a_n.

References

[1] E. Bombieri, Le grand crible dans la théorie analytique des nombres. *Astérisque* **18** (Société Mathématique de France, 1974).

[2] E. Bombieri, J. Friedlander and H. Iwaniec, Primes in arithmetic progressions to large moduli. *Acta Math.*, **156** (1986), 203–51.

[3] E. Bombieri, J. Friedlander and H. Iwaniec, Primes in arithmetic progressions to large moduli, II, *Math. Ann.*, **277**: 3 (1987), 361–93.

[4] V. Brun, Über das Goldbachsche Gesetz und die Anzahl der Primzahlpaare. *Archiv for Math. og Naturvid.*, **B34**: 8 (1915), 19 pages.

[5] V. Brun, Le crible d'Eratostène et le théorème de Goldbach. *Videnskaps. Skr., Mat.-Naturv. Kl. Kristiana*, no. 3 (1920), 36 pp.

[6] D. Bump, *Automorphic Forms and Representations*, (Cambridge: Cambridge University Press, 1997).

[7] A. C. Cojocaru, E. Fouvry and M. Ram Murty, The square sieve and the Lang–Trotter conjecture. *Can. J. Math.* (to appear).

[8] H. Davenport, *Multiplicative Number Theory*, 3rd edn (New York: Springer Verlag, 2000).

[9] J.-M. Deshouillers and H. Iwaniec, Kloosterman sums and Fourier coefficients of cups forms. *Inv. Math.*, 70 (1982), 219–88.

[10] W. Duke and E. Kowalski, A problem of Linnik for elliptic curves and mean value estimates for automorphic representations. *Inv. Math.*, **139** (2000), 1–39.

[11] P. D. T. A. Elliott, *Probabilistic Number Theory I: Mean Value Theorems*, New York, Berlin: Springer-Verlag, 1980.

[12] P. D. T. A. Elliott, *Probabilistic Number Theory II: Central Limit Theorems*, New York, Berlin: Springer-Verlag, 1980.

[13] P. D. T. A. Elliott and H. Halberstam, A conjecture in prime number theory. *Symp. Math.*, **4** (1968/9), pp. 59–72.

[14] C. Elsholtz, Some remarks on the additive structure of the set of primes. In *Number Theory for the Millennium I*, (Natick, MA: A. K. Peters, 2002), pp. 419–27.

[15] V. Ennola, On numbers with small prime divisors. *Ann. Acad. Sci. Fenn. Ser. AI*, **440** (1969), 16 pp.

[16] P. Erdös and C. Pomerance, On the normal number of prime factors of $\phi(n)$. *Rocky Mountain J.*, **15** (1985), 343–52.

[17] Jody Esmonde and M. Ram Murty, *Problems in Algebraic Number theory*, (New York: Springer-Verlag, 1999).

218

[18] E. Fouvry, Autour du théorème de Bombieri-Vinogradov. *Acta Math.*, **152**: 3–4 (1984), 219–44.

[19] E. Fouvry, M. Nair and G. Tenenbaum, L'ensemble exceptionnel dans la conjecture de Szpiro. *Bull. Soc. Math. France*, **120**: 4 (1992), 483–506.

[20] P. X. Gallagher, The large sieve and probabilistic Galois theory. *Analytic Number Theory, Proc. Symp. Pure Math.*, **24**, (AMS, 1973), 91–101.

[21] P. X. Gallagher, A larger sieve. *Acta Arithm.*, **18** (1971), 77–81.

[22] S. Gelbart, *Automorphic Forms on Adele Groups* (Princeton, NJ: Princeton University Press, 1975).

[23] R. Gupta and M. Ram Murty, A remark on Artin's conjecture. *Inventiones Math.*, **78** (1984), 127–30.

[24] D. R. Heath-Brown, The square sieve and consecutive squarefree numbers. *Math. Ann.*, **266** (1984), 251–9.

[25] D. R. Heath-Brown, Artin's conjecture for primitive roots. *Q. J. Math., Oxford*, **37**: 2 (1986), 27–38.

[26] L. K. Hua, *Introduction to Number Theory* (Berlin: Springer-Verlag, 1982).

[27] H. Halberstam and H. E. Richert, *Sieve Methods*, London Mathematical Society monographs, No 4, (London, New York: Academic Press, 1974).

[28] G. H. Hardy and E. M. Wright, *An Introduction to the Theory of Numbers* (Oxford, 1960).

[29] M. Harper and M. Ram Murty, Euclidean rings of integers. *Can. J. Math.*, **56** (2004), 71–6.

[30] M. Hindry and J. Silverman, *Introduction to Diophantine Geometry* (New York: Springer-Verlag, 2001).

[31] C. Hooley, *Applications of Sieve Methods*, (Cambridge: Cambridge University Press, 1976).

[32] K. Ireland and M. Rosen, *A Classical Introduction to Modern Number Theory*, 2nd edn (New York: Springer-Verlag, 1998).

[33] H. Iwaniec, Fourier coefficients of cups forms and the Riemann zeta function. *Sém. Th. Nb. Bordeaux*, exposé no. 18, (1979–1980) 36 pp.

[34] H. Iwaniec, Rosser's sieve. *Acta Arith.*, **36** (1980), 171–202.

[35] W. Li, *Number Theory with Applications* (Singapore: World Scientific, 1996).

[36] Yu. V. Linnik, The large sieve. *Dokl. Akad. Nauk. SSSR*, **30** (1941), 292–4.

[37] Yu-Ru Liu and M. Ram Murty, The Turán sieve and some of its applications, *J. Ramanujan Math. Soc.*, **13**:2 (1999), pp. 35–49.

[38] Yu-Ru Liu and M. Ram Murty, A weighted Turán sieve method, to appear.

[39] J. Merlin, Sur quelques théorèmes d'Arithmetique et un énoncé qui les contient. *C.R. Acad. Sci.Paris*, **153** (1911), 516–18.

[40] J. Merlin, Un travail de Jean Merlin sur les nombres premiers. *Bull. Sci. Math.*, **39**:2 (1915), 121–36.

[41] H. Montgomery, *Topics in Multiplicative Number Theory*, (New York: Springer Verlag, 1971).

[42] H. L. Montgomery and R. C. Vaughan, The large sieve. *Mathematika*, **20** (1973), 119–34.

[43] Y. Motohashi, An induction principle for the generalizations of Bombieri's prime number theorem. *Proc. Japan Acad.*, **52**:6 (1976), 273–75.

[44] M. Ram Murty, Artin's conjecture for primitive roots, *Math. Intelligencer*, **10**:4, (1988), 59–67.

[45] M. Ram Murty, *Problems in Analytic Number Theory*, (New York: Springer-Verlag, 2001).

References

[46] M. Ram Murty, Sieving using Dirichlet series. In *Current Trends in Number Theory*, ed. S. D. Adhikari, S. A. Katre and B. Ramakrishnan. (New Delhi: Hindustan Book Agency, 2002), pp. 111–24.

[47] M.Ram Murty, Sieve methods, Siegel zeros and Sarvadaman Chowla. In *Connected at Infinity, Texts Read. Math.*, **25** (2003), pp. 18–35.

[48] M. Ram Murty and V. Kumar Murty, Prime divisors of Fourier coefficients of modular forms. *Duke Math. Journal*, **51** (1984), 57–76.

[49] M. Ram Murty and V. Kumar Murty, An analogue of the Erdös–Kac theorem for Fourier coefficients of modular forms. *Indian J. Pure Applied Math.*, **15**:10 (1984), 1090–101.

[50] M. Ram Murty and F. Saidak, Non-abelian generalizations of the Erdös–Kac theorem. *Can. J. Math.*, **56** (2004), 356–72.

[51] M. Ram Murty and N. Saradha, On the sieve of Eratosthenes. *Can. J. Math.*, **39**:5 (1987), 1107–122.

[52] M. Ram Murty and N. Saradha, An asymptotic formula by a method of Selberg. *C.R. Math. Rep. Acad. Sci. (Canada)*, **XV**:6 (1993), 273–77.

[53] B. M. Nathanson, *Additive Number Theory. The Classical Bases* (New York: Springer-Verlag, 1996).

[54] D. A. Raikov, Generalisation of a theorem of Ikehara–Landau (in Russian). *Mat. Sbornik*, **45** (1938), 559–68.

[55] S. Ramanujan, A proof of Bertrand's postulate. *J. Indian Math. Soc.*, **11** (1919), 181–2.

[56] R. Rankin, Contributions to the theory of Ramanujan's function $\tau(n)$ and similar arithmetical functions, III – A note on the sum of the Fourier coefficients of integral modular forms. *Proc. Camb. Phil. Soc.*, **36** (1940), 150–1.

[57] A. Selberg, Bemerkungen über eine Dirichletsche Reihe, die mit der Theorie der Modulformen nahe verbunden ist. *Archiv for Mathematik og Naturvidenskab B.*, **43**:4 (1940), 47–50 (see also A. Selberg, *Collected Papers*, vol. 1, pp. 38–41, Springer-Verlag, 1989).

[58] A. Selberg, An elementary proof of the prime number theorem. *Ann. Math.*, **50** (1949), 305–13.

[59] A. Selberg, The general sieve method and its place in prime number theory, *Proc. Int. Congress of Mathematicians* Cambridge, MA (Providence, RI. American Mathematical Society, 1950), vol. 1, pp. 289–262.

[60] A. Selberg, Lectures on sieves, *Collected Papers*, vol. 2. (Berlin: Springer-Verlag, 1991), pp. 66–247.

[61] J.-P. Serre, *A Course in Arithmetic*, (Springer, 1973).

[62] J.-P. Serre, Majorations de sommes exponentielles. *Journées Arith. Caen, Astérisque*, **41–42** (1977), 111–26.

[63] J.-P. Serre, Divisibilité de certaines fonctions arithmétiques', *L'Ens. Math.*, **22** (1976), 227–60 (see also *Collected Papers*, vol. 3, pp. 250–83, Springer-Verlag, 1986).

[64] J.-P. Serre, Quelques applications du théorème de densité de Chebotarev. *Pub. Math. I.H.E.S.*, **54** (1981), 123–201.

[65] J.-P. Serre, Spécialisation des éléments de $Br_2(\mathbb{Q}(T_1, \dots, T_n))$. *C. R. Acad. Sci., Paris*, **311** (1990), 397–422 (see also *Collected Papers*, vol. 4, pp. 194–199 Springs-Verlag, 1986).

[66] G. Shimura, *Introduction to the Arithmetic Theory of Automorphic Functions*, (Princeton, NJ: Iwanami Shoten and Princeton University Press, 1971).

[67] C. L. Siegel, The integer solutions of the equation $y^2 = ax^n + bx^{n-1} + \cdots + k$. *J. London Math. Soc.*, **1** (1920), 66–8.

[68] G. Tenenbaum, *Introduction to Analytic and Probabilistic Number Theory* (Cambridge: Cambridge University Press, 1995).

[69] E.C. Titchmarsh, A divisor problem. *Rend. Circ. Mat. Palermo*, **54** (1930), 414–29.

[70] R. C. Vaughan, A note on Schnirelman's approach to Goldbach's problem. *Bull. London Math. Soc.*, **8**:3 (1976), 245–250.

[71] R. C. Vaughan, On the estimation of Schnirelman's constant. *J. reine Angew. Math.*, **290** (1977), 93–108.

[72] P. Weinberger, On Euclidean rings of algebraic integers. *Proc. Symp. Pure Math.*, **24** (1973), 321–32.

[73] A. Wiles, Modular elliptic curves and Fermat's last theorem. *Ann. Math.* **141**:2 (1995), 443–551.

[74] R. Wilson, The Selberg sieve for a lattice. In *Combinatorial Theory and its Applications, III*, Proc. Colloq. Ballatonfüred, 1969 (Amsterdam: North-Holland, 1970), pp. 1141–9.

[75] A. Wintner, On the prime number theorem. *Amer. J. Math*, **64** (1942), 320–26.

Index

O-notation, 1
$\Omega(\cdot)$, 43, 59, 110
$\Phi(x, z)$, 63, 68, 69, 76, 78, 79, 82, 85, 113, 117, 133
$\Psi(x, y)$, 29, 30, 60, 68, 69, 76, 78, 79
$\asymp$-notation, 2
$\gg$-notation, 1
$\ll$-notation, 1
$\nu(\cdot)$, 2, 32–35, 37, 43, 44, 59, 109
$\pi(x)$, 5, 11, 12, 59, 62, 64, 69, 76, 79, 85, 110, 113, 117
$\pi(x; k, a)$, 38, 76, 109, 112, 124, 125, 130, 133, 147, 152, 167
$\pi_K(x)$, 36
$\psi(x)$, 7, 11, 12, 148
$\psi(x, \chi)$, 147, 152
$\psi(x; k, a)$, 147, 152
$\sim$-notation, 2
$\theta(x)$, 5, 11
$d(\cdot)$, 10, 11, 51, 58, 59, 109, 157, 172, 174, 175, 188, 189
o-notation, 2

additive arithmetical function, 45
analytic density, 124, 130
arithmetical function, 13
Artin, 195
Artin L-function, 43
Artin's conjecture on primitive roots, 172, 195

Balasubramanian, 105
Barban-Davenport-Halberstam theorem, 147
Bertrand's postulate, 6, 12
Bombieri, 135, 156, 167, 177, 178

Bombieri-Vinogradov theorem, 39, 40, 42, 44, 125, 135, 156, 157, 167, 171, 173, 175, 191, 195
Brun, 15, 73, 80, 88, 90, 177
Brun's pure sieve, 63, 81, 82, 86, 113
Brun's sieve, 3, 73, 93, 98, 100, 102, 109–112, 127, 129, 201
Brun's theorem, 73, 74
Brun-Titchmarsh theorem, 109, 125, 127, 174
Buchstab's function, 78
Buchstab's identity, 29, 30, 78, 79, 89
Buniakowski's conjecture, 38, 124
Bushnell, 215

Cauchy-Schwarz inequality, 18, 27, 45, 46, 104, 106, 112, 141, 144, 146, 152, 154, 162, 164, 174, 191, 202, 206
Chebotarev density theorem, 58
Chebycheff, 1, 5, 9, 113
Chebycheff's theorem, 6, 10, 12, 38, 52, 59, 70, 76
Chowla, 172
completely multiplicative function, 76, 123
cyclotomic polynomial, 19, 28, 29

Davenport, 135
de la Vallée Poussin, 6
de Polignac's conjecture, 106
Dedekind's theorem, 36
Deligne, 23
Deshouillers, 105, 172
Dickman, 30
Dickman's function, 29, 30
Dirichlet, 124
Dirichlet character, 142, 151, 204, 207

222

Printed in the United States
By Bookmasters